SASQUATCH: BIGFOOT

The Continuing Mystery

Thomas N. Steenburg

ISBN 0-88839-312-1

Revised printing 1993

Cataloging in Publication Data
Steenburg, Thomas N. (Thomas Nelson)
Sasquatch: Bigfoot: the continuing mystery

First ed. has title: The Sasquatch in Alberta.
ISBN 0-88839-312-1

1. Sasquatch—Alberta. 2. Sasquatch—British Columbia. I. Title. II. Title: The Sasquatch in Alberta.
QL89.2.S2S84 1993 001.9'44 C93-091370-1

Printed in Hong Kong

Cover design: Betty A. Janner

Published simultaneously in Canada and the United States by

HANCOCK HOUSE PUBLISHERS LTD.
19313 Zero Avenue, Surrey, B.C. V4P 1M7
(604) 583-1114 Fax (604) 538-2262

HANCOCK HOUSE PUBLISHERS
1431 Harrison Avenue, Box 959, Blaine, WA 98231-0959
(206) 354-6953 Fax 538-2262

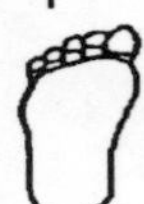

CONTENTS

To Barbara,
you've given me the will to carry on
with my research.

Anyone who has information concerning the Sasquatch is invited to call to me:

Thomas Steenburg
Calgary, Alberta

I'm in the book.

FOREWORD

Tom Steenburg recently wrote *The Sasquatch in Alberta.* In this book he discussed the possibilities that the Sasquatch may exist in Alberta. This came as a surprise to many people who always associated the stories about Sasquatch with the mountains of British Columbia and with the region of the Harrison Lake area particularly.

In the present book, Steenburg has extended his research on Sasquatch into British Columbia. He describes visits to isolated places and small towns, asking questions about the Sasquatch even when people laugh at him, including R.C.M.P. officers. He interviews those who claim to have seen a Sasquatch and tries to make a sense of all these stories.

Sasquatch, judging by what people believe to have seen, emerges as a pretty uniform being. The creature is straight and very tall; at least 7 feet in height, perhaps 8 feet on average. The problem is, of course, its sex - people claim to have seen it, none of the reports mentions the gender. Among the higher primates, the males are always taller and larger than the females. Whatever the tallness of the Sasquatch, the fact that it walks on two legs, not four, is a sure sign that it is anatomically closer to a human being than to an ape. We call such beings hominids.

The stride that Sasquatch makes, judging from the imprints discovered in soil, is said to be up to 6 feet long while the foot should be at least 13 inches in length. The foot looks much like an enlarged human foot and not at all like the foot of an ape. The case of the arms is somewhat strange. They are described in one instance as ending below the knees, which is actually an ape characteristic and not at all human. Is it perhaps possible, that legs became human-like while the arms kept some ape-like characteristics?

The body is described as being completely or almost completely covered with fur. The parts of the body where it is said that there is no hair are the face and palms of the hands. This is consistent with earlier descriptions of the Sasquatch in other parts of North America. The color of the fur is said by various witnesses to be black, brownish-black, brownish or silvery. This variation of colors makes sense. The young ones have black hair and when they get old their fur becomes silvery. The same way that humans in their old age become grey. As for the brownish color of the fur it was suggested that this develops from eating various berries: the juice sticks to the fur making it reddish or brownish.

Another element to discuss, is that Sasquatches soften smell, and sometimes of sulphur. That Sasquatches smell bad is mentioned often in many previous reports. Animals do not usually smell but it is common among humans and has something to do with the sweat glands in the skin. This fact again suggests that Sasquatches are anatomically closer to humans than to apes. But that they smell of sulphur is very puzzling. I do not know the reason for this, but it has to be connected to some glands in the skin. It was believed in medieval Europe that the devil always smelled of sulphur and sulphur became associated with hell and its fire. Could it have been that people in medieval Europe seeing some kind of standing hairy beings smelling of sulphur were actually seeing a hominid. Stories about some non-human hominids are mentioned in Europe and all around the world.

Steenburg's book is full of such interesting details that make it worthwhile to read. Anybody interested in pursuing the Sasquatch can learn a great deal from it, how to interview witnesses too.

VLADIMIR MARKOTIC

INTRODUCTION

Was it merely an Indian Legend told to early explorers, a story characteristic of the native's mythical culture? Or, were the stories of giant hairy man-like apes actual reports of an animal that has managed to mystify its researchers and elude Western civilization for nearly three hundred years?

While the European pioneer travelled across the continent of North America he was entertained by its colorful natives and their flashy heroic tales of this elusive man-like beast. The names for this animal were as varied as the tribes of the North American Indians, themselves. Sasquatch (English pronunciation) was only one such name.

The same cultural differences that had entertained the settler, however, were to become part of the settlers' intolerance of the Indian ways. When the natives refused to give up their stories, religious beliefs and land and join the white man's "more civilized approach" to life their cultural structure was first discredited and then destroyed. The Indian was eventually cast away onto "reserved land" along with his traditions, lifestyles and legends.

Long after early explorers had laid claim to and secured their lands and cities, however, the descendants of the settlers began moving back to unpopulated territories to enjoy the beauty of the wild mountains and attempt to preserve what was left of the forests. It was then that the Legend of the Sasquatch was reborn.

Campers, hunters, adventurists were returning home with incredible reports of one or more giant hair-covered man-like beasts, huge man-like footprints and strange unidentified animal calls. Only this time, the "civilized man" would not hear; no such animal had ever been seen publicly or found by the scientific community. The Sasquatch became a joke, a myth, a money making event, a mere speculation to all but a select few men whose determination to prove its existence would not be quenched.

Several years ago, two such men grabbed the attention of the world with a film they had made of the elusive creature. Try as it might, the scientific community could find no evidence to excuse the film as a hoax or a set-up. As legitimate explorers, researchers and a few scientists joined in on the hunt to find the Sasquatch, the race for evidence was on; to prove the existence of and settle the Sasquatch mystery once and for all.

John Green, a Canadian widely renowned for his research into this subject proposed the theory that the Sasquatch is an ape, a great ape residing within but not exclusive to the North American continent. Professor Grover Krantz, an anthropologist from Washington State University has, from his studies, come to the very same conclusion. From his research into fossil records, Professor Krantz has surmised that the Sasquatch could be a close descendent of the Gigantopithecus. These, however, are merely theories and until the animal itself is proven to exist, all theories remain just theories.

In today's "enlightened" society man finds it quite easy to believe in the ghosts he can never see, the unscientific and possibly ludicrous ramblings of strange men with foreign beliefs but has difficulty believing what his fellow man swears to have seen, touched, heard or known. This type of objectivity regarding the Sasquatch, however, is not only necessary, but of extreme importance. In order to prove the animal is alive and is indeed an animal, no room can be left for frivolous acceptance. Such would hurt the progress of research and open the door to classifying Sasquatch with mythical creatures before proof can be found.

There can be no exact number showing how often this animal has been seen as it is quite possible that most sightings go unreported. Because the Sasquatch has been regarded as a "silly myth" for so long now, it is strongly believed that many witnesses will clam-up to fend off negative publicity and ridicule. However, at the time of writing, 349 reported incidents have been brought to my attention.

In the United States there has been 57 reported sightings in Oregon, 98 reported sightings in Washington and 66 reported sightings in Northern California. In Canada there has been 105 reported sightings in British Columbia (B.C.) and 23 reported sightings in Alberta. According to the numbers in my files, the Sasquatch, if it indeed exists, seems to be dwelling in the forests, mountains and costal regions of Western North America.

In Alberta, a small part of the public school curriculum deals with "modern unproven monsters". The Ogopogo, The Loch Ness Monster, The Abominable Snowman and the Sasquatch. The Sasquatch seems to be presented as a single monster with a "boogy-man" type personality and the sighting of such equalized with that of a vampire, werewolf or a witch on a broomstick. The possibility of its existence as a rare, yet unfounded animal possibly harmless to the human race has all but been excused. This attitude has helped to place the animal along side fairy tale figures most children leave behind in their adulthood. Unfortunately, it has also led most adults to believe that there is no evidence of the animal and no serious research into the subject - What man in his right mind would spend time researching Peter Pan, Cinderella, The Wicked Witch of the West or, and especially, the "boogy-man"!

Common questions asked by people who discover that serious research is taking place are: "If it is in the mountains, why haven't I seen one? Why hasn't one been hit with a car? There are bear sightings reported and dealt with all the time, why hasn't the Sasquatch been given the same treatment?". One of the most common questions is,"Why hasn't the bones or remains of a dead Sasquatch been found?" This is the toughest question to answer.

However, finding a Sasquatch in the mountains or forests would be like looking for a needle in a haystack. If it is there, the chances of finding it are quite rare. Professor Grover Krantz estimated some years ago that there may be only about 200 Sasquatches living in the Pacific Northwest.

The province of Alberta is larger than many countries. Much of it, however, is unpopulated mountain and forest terrain. Tourists, campers and hunters do run into bears and other wildlife but not everyday. Using the bear as an example, we will examine the possibility of running into a Sasquatch in the wilderness.

Manhunts, plane searches, war games, nature excursions and full tours take place in the foothills and Rocky Mountains of Alberta regularly. During these times, bears are rarely sighted. The bear population of Alberta is quite high, yet a person who lives and travels through Alberta for ten to twenty years may never see one unless they go to a zoo displaying the bears. If the Sasquatch does exist, it is obviously a rare animal who may avoid or fear human contact. If this be the case then 349 reported incidents involving the Sasquatch is not an unbelievably high number.

Of the sightings reported to me, each person involved was questioned extensively. If they had enough information to be convincing an interview would be granted to get to the facts and expose the fallacies. Many of these people have come forward in fear of exposure. They are convinced that they have seen a Sasquatch but are tired or afraid of ridicule. This is their story.

CHAPTER ONE
"Bigfoot And The Big Horn Dam"

On (July 4, 1986) I placed an advertisement in the classified section of the Calgary Herald requesting information on the Sasquatch. For years having researched, read, studied the subject of this animal and having a keen interest in the proving of its existence, the ad was placed so that recent sightings in Alberta could be reported to someone and the story weighed for legitimacy.

Early afternoon of January 25, 1988 the phone rang. Debra Malone of Calgary, Alberta, was ready to let out a secret she held inside for sixteen years - she had seen, up close, an animal believed by her to fit the description of the Sasquatch. We set up a meeting.

Although I had many prank calls as a result of the ad, Debra quickly made it clear that she was determinedly serious and would not put up with being doubted. She was an outspoken friendly young woman who could not be classified as flakey in any way. She had, in spite of her own feelings, kept the Sasquatch sighting of her early teens relatively quiet. Those she had told interpreted the story as frivolous gesturing and conversational rambling. Her husband, who was possibly the closest to believing her, viewed the tale with skepticism.

Emotionally and often close to tears, Debra related what had happened in the summer of 1972 north of Abraham Lake on Highway 11 near the then in construction site of the Big Horn Dam in Alberta.

Debra, her younger brother Allen and her parents set out on the weekend to a secluded area near Nordegg and by Abraham Lake, just off of forestry road. During the drive they had passed the construction site of the Big Horn Dam. Past that point, however, the land was unpopulated, unexplored and undamaged by civilian occupation.

They continued driving on the highway for nearly 10 miles before turning into the bush. Debra recalls going about 50 miles further before stopping to set up camp. The area was dense with trees and berry bushes.

The day was still young and the family had made no specific plans. Debra was bored already. Instinctively attracted to nature though, she started towards the tree line and decided that now would be an opportune time for an exploratory hike in the woods.

She wandered through the natural mazes of tall pine and wild flowers taking in the freshness of the air and the untouched beauty. Eventually her wandering brought her to a bush of ripe berries and Debra bent down to pick them from the vines.

It was then that the stench hit her. An unfamiliar odor filled her nostrils and it was unlike any smell she had encountered before. Debra describes it as "...not as bitter as a skunk smell but muskier than that of a dog in heat". At that very same moment, the young city girl felt the presence of someone, something watching her. She whirled around, half standing, half squatting and found herself facing a creature who resembled no other creature she had ever seen or heard about.

Not more than 50 feet away in clear view and standing next to a thick line of bushes was an estimated eight foot tall hairy animal poised quite naturally in a perfectly human upright pose. *His arms hung down past his knees, his shoulders were at least three feet across and he was covered with long, matted, dirty brown and grey hair. The creature's nose jutted out from its face like a flattened ball of putty but it was his eyes which immediately claimed her attention - they were slate black, scrutinizing, even scary.

"That's no moose!" was Debra's first thought. She knew it couldn't be a bear... so just what was it?! Turning around and standing up her eyes met his. According to Debra it appeared so completely at home, so much in its natural

---

* *For the sake of grammatical clarity the male gender will be used in reference to the Sasquatch were no gender is evident.*

element that the possibility of it being otherwise didn't even cross her mind. He was standing up just like a person would but it was an animal. There were no solid trees in the immediate area, nothing to lean on or hold himself up by. He was not engaging in any activity, in fact, he just stood there staring right back at Debra.

For nearly ten minutes she watched him and he watched her right back. The creature made no motion of attack or sign of fear or even curiosity. He did not move a muscle and appeared to be undaunted by her presence. His eyes, she claims, communicated wisdom and instinct.

After what seemed like forever to Debra, the animal let out a deep throaty moan. Nearing a state of shock, she turned and ran, fighting her way through the trees, trying to remember the way she came; a wave of nausea hit, Debra started to vomit.

Finally finding her way back to camp, she burst through the tree line smelling of vomit, tears rolling down her cheeks. Her face went pale, she shook all over and was barely able to walk. Finding her brother she tried to explain what had happened.

Debra's mother heard the story and was convinced that her daughter had been spooked by a wild animal, possibly a bear. She listened, but did not believe that the girl had seen what was described - a description which fit that of a legendary mythical creature known as the Sasquatch.

Debra slept with her mom that night, still shaken and possibly in shock. Later, she was scolded for holding fast the story of "the big hairy man in the woods".

Not having previously known what a Sasquatch was, or having any former knowledge of the possible existence of such an animal, the sighting opened the door for Debra's education on "unproven animals". Approximately a year later she learned through a school assignment that such an animal is claimed by some to be in existence in North America. For the purpose of letting her story out, without facing more ridicule and scolding, she did a major writing assignment and presentation on book evidence and gathered knowledge of the Sasquatch. After the presentation, her instructor commented that: "it was written as if she had seen one herself".

Growing up, Debra would mention the Sasquatch sighting from time to time. In a circle of friends when the conversation turned to the unknown at a party or with a group of people she was close to, she would make the short and simple statement "...I've seen a Sasquatch".

She never gave a detailed account though because she was afraid of being laughed at.

Before she was married, however, Debra decided to tell her fiance about the experience. He, like most, was very reluctant to believe that the creature she encountered in 1972 was the same creature of myth classified with UFO's and unicorns. But, knowing her as he did it could not be denied that she had seen something and that she believed it was a Sasquatch.

Although Debra's story was told less and less she could not put it out of her mind. Eventually the experience invaded her dreams at night causing nightmares in which she would recall the experience as if it had happened that very day.

When at last, in 1988, Debra came across my ad in the Calgary Herald requesting information on the Sasquatch, she did not want to phone. It had taken some time for her to lift that receiver and dial the number but in doing so, a new chapter of her life would begin - she would not be alone in her experience. Others had too seen what she had seen.

In the course of my interview with her she admitted to wanting to see it again - only this time with someone else who could be a witness that the creature really did exist.

Is it possible that Debra Malone made up this story? That she carefully devised the tale in order to meet with conversational needs or have some claim to fame? Yes, such a thing is possible and has actually happened. The facts which substantiate her truthfulness, however, cannot be ignored.

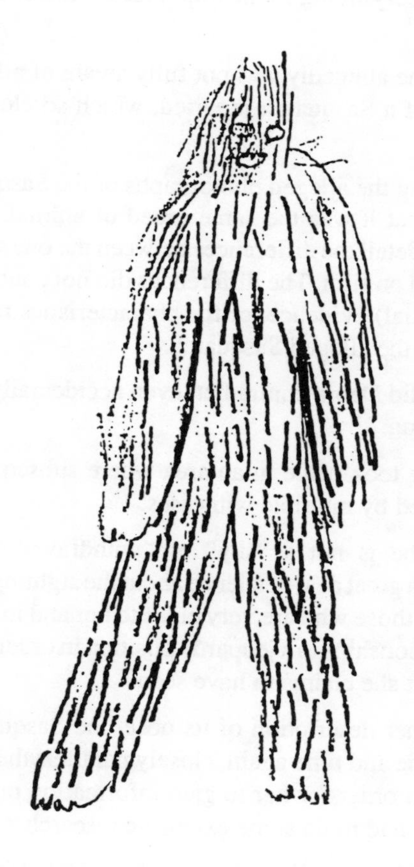

Picture Drawn by Debra Malone, Age 13

- Prior to 1972, Debra had no knowledge of the existence of an unproven animal allegedly living in the unpopulated forest areas of the Pacific Northwest.
- Debra Malone allegedly was not fully aware of all the physical characteristics, of a Sasquatch reported, which so closely paralleled her description.
- Upon viewing the alleged photographs of the Sasquatch with me, she confirmed that it was the same breed of animal, then proceeded to describe, in detail the differences between the one she had seen and the one captured on film. The differences did not contradict any gathered (circumstantial) evidence, or the characteristics believed to be common among the alleged Sasquatch.
- At no time did Debra contradict, even accidentally, her original story of the sighting.
- Her sighting took place in an area where subsequent sightings have been reported by multiple witnesses.
- Although she is not a "shy" or "withdrawn" person, Debra has maintained a great deal of silence about the sighting for fear of ridicule, but has told those who are very close to her and in doing so has placed certain relationships in a jeopardy position in order to maintain that she did see what she claims to have seen.
- Excluding her description of its nose, the Sasquatch, as it has been reported time and time again, closely parallels the description she has provided. In order for her to gain information on these features, she would have had to do some extensive research.
- The stench of the animal is not a usually reported factor, when sightings are published, but the smell is an uncommon factor in most sightings and Debra's description of such matches both previous and recent descriptions.

A considerable amount of time has passed since my interview with Debra. Presently, she is collecting information on the Sasquatch, reading everything she can find about the subject and taking in every documentary. She is no longer afraid to "relate" the experience and now holds the attitude that if someone does not believe her, that is their choice. Her story, related and re-related has not changed a bit since the original interview. Her willingness to share the story, she claims, is a result of knowing that others right here in Alberta and more specifically in the area of the Big Horn Dam have also reported sighting a Sasquatch.

Her hope is that the animal can one day be brought, without harm, into captivity and documented as a legitimate species. She hopes that her husband will have the opportunity to see one and know without apprehension that her story is not a just a magnificent spun tale.

The Big Horn Dam, close to where Debra Malone claims to have sighted a Sasquatch, is the site of the most publicized Sasquatch incident ever reported in Alberta.

Just past Banff National Park and West of Rocky Mountain House, by Highway 11 is the Big Horn Dam. In the summer of 1969 the dam was yet a construction site. Much of the surrounding area was untouched by modern man. Aside from the few men working in the area there appeared no evidence of human occupation in the immediate area - the land was considered "virgin land".

On August 23, 1969 a five man construction team working on the water pumping station encountered a Sasquatch. Harley and Stan Peterson of Condor, Floyd Engel of Eckville, Guy L'Heureux of Rocky Mountain House and Dale Boddy of Ponoka Alberta, noticed a distant figure on the high riverbank. They gathered together to watch it. According to Harley it looked like an incredibly large man. He described it as enormous, head slightly bent forward and very hefty.

Dale Boddy reported that it was too tall and it's legs too thin for a bear. As well, the speed it was moving at - approximate strides of 6 feet in length

- prompted him to believe that the creature was too large for any animal he had heard or known of.

Neither he nor the others with him had a pair of binoculars or a camera. They all turned to watch the figure. For nearly a half an hour the distant stranger stood stationary. Finally it sat down. Ten minutes later the thing stood up again hesitated for awhile then walked along the ridge of the riverbank and into the tree line. That is when the men lost sight of it. The animal had been in view for nearly 45 minutes.

Astonished by its apparent height and curious to know just how tall it was, two of the men headed to the riverbank to do a size check. When they arrived at the exact location, the three who stayed behind plainly realized that the creature who just walked off was more than twice the height of the men who just arrived, which would make it nearly 15 feet tall.

When the press got ahold of the story, witnesses to the Sasquatch animal came pouring out of the wood work. Mark Yellowbird, a Cree Indian also working on the dam had previously entertained his fellow workers with stories of human footprints along Rabbit Creek nearly 14 1/2 inches long and spanning more than six feet apart.

According to a Calgary Herald report August 30, 1979 (Bob Hage), Mark Yellowbird's daughter Edith along with three middle aged women had sighted "...four of the furry beasts working at something part way up a mountain along the David Thompson Highway".

Alec Shortneck, a worker at the dam, was clearing bush for construction when he reported to have sighted a strange creature watching him. He saw the figure about 50 yards away and was too stunned to do or say anything about it. Alec went on chopping at the bush and the figure left.

Vern Saddleback, a native in the area, claims that many people have actually seen the tracks yards away from his camp but never reported them or took the matter seriously. One Indian group, however, did report sightings. A band of Indians led by Chief Joe Smallboys were travelling in the area between Banff National Park and Nordegg in the spring and summer of 1969.

It was during this time many members of his band allegedly saw up-close, the Sasquatch.

In the opinion of the residents of surrounding towns and people on the local Indian reservations, a band of animals fitting the description of the elusive Sasquatch are roaming the area outside of Nordegg and near the Big Horn Dam sight.

The ridge at the Big Horn Dam in which workmen reported seeing a Sasquatch in 1969.

The late Chief Walking Eagle shared the opinion. He was willing to talk about it to his friends but never outsiders. According to The Edmonton Journal, August 30, 1969 (Nick Lees) the Chief was sure that his opinion would not be regarded as serious and he would thus be laughed at. So, he kept his ideas to himself and if he had any evidence of the existence of the Sasquatch, the knowledge was buried with him.

When the story reached full coverage Mr George Harris, a retired bulk fuel businessman, began making plans for an expedition to the Big Horn Dam site. Mr. Harris had, at one time, taken photographs of enormous footprints in the sand near the dam site which he believed were Sasquatch prints. In his report he stated that the prints could have belonged to a male and female of the species because of their varying sizes (13 and 17 inches long).

Gunter Schung, a guide and trapper from Nordegg, also made plans. He agreed to accompany Mr. Harris on the expedition. The two men were determined to track the creature down.

As would be expected, though, the expedition never took place.

During the press heyday with the Big Horn Dam sighting, a new angle was cast on the Sasquatch question: is the animal a revival of an Indian folklore legend used to scare the white man off from areas considered "sacred" by local bands of natives?

At the time of the sighting the Wesley Band of Indians were boycotting employment opportunities at the dam in a protest against the structure. The band's Chief John Snow spoke that his people felt that the dam would erase a sacred burial site and endanger his reserve located six miles downstream. Abraham Lake, which was formed by the dam, would undoubtedly ruin some of the land of the Kootenay Plains, destroy trap lines and game in the area.

It was because of these ideas that a journalist for the Edmonton Journal reported the possibility of Indian magic used to create some illusion in order to scare workers off the sight and possibly cancel out a ratifyed bill between the Provincial Government and Calgary Power Ltd. which solidifyed the construction of the dam.

It is possible, however, that all five men devised the story amongst themselves and set out to draw the attention of the media. Yet another possibility to be considered is "angles and the sun playing tricks with and obstructing a proper and clear view of an object or thing which could have been in motion at the sighting time".

The first possibility cannot be absolutely proven or disproven at this point. If the men were called together again and interviewed, first separately and then as a group, the time factor involved (20 years) would invalidate any changes or contradictions as it may be difficult to clearly remember all details 20 years after the fact.

Still, it is also possible that the story had not been true even if all five men remained in agreement to this date. If the men had lied to the media, the chances of coming clean now are quite slim. Also, it is relatively acceptable to believe that made-up stories are just as easy to remember as are stories of fact. Consequently, if this was not a true sighting it is twenty years too late to prove it.

Would it have been possible, though, for the sun to have hit such an angle of the earth at a tree or moving animal and cast a dark shadow or cause something to appear to be fifteen feet high? Yes, there are reported cases of the sun playing such a deception on man. In this case, however, it would be impossible. First there were five witnesses. Each of the men would have been standing at a slightly different angle than the other, thus one would be unable to see from the exact vantage point that another man was viewing the creature from. Secondly, all five men attested that the creature stayed long enough for them to observe it standing up, sitting down, standing up again and walking into the tree line. Evidently, the sun cannot accomplish a "trick angle" for the amount of time the men claimed to have watched the creature, nor could the sun cause such a trick to move across the ground and into the tree line.

Obviously, in 1969 at the Big Horn Dam construction site five men saw a creature standing, sitting and moving. The creature was over six feet tall and possibly as high as 15 feet tall. Was it a Sasquatch or a manifestation of Indian magic?

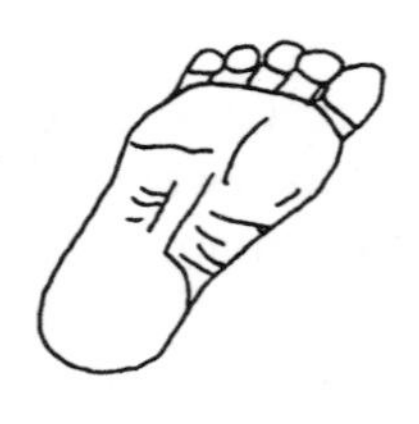

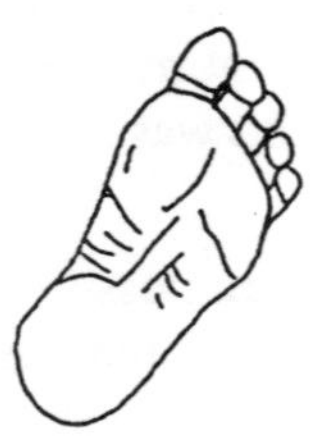

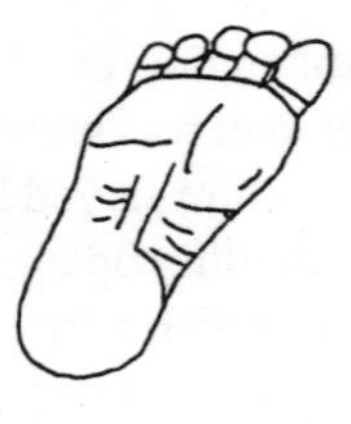

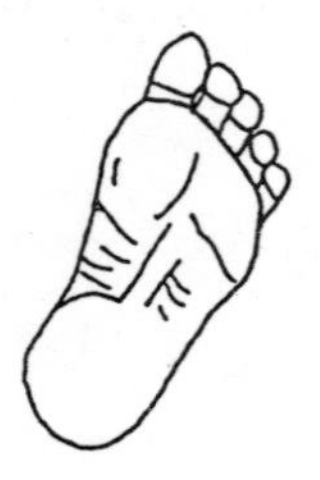

CHAPTER TWO
"Needle In A Haystack"

The most common question asked of researchers is why, if the Sasquatch exists, hasn't the live or dead body or parts thereof been discovered? There are thousands of known animals in the world today, both rare and common, observed daily around the globe. The very idea of an animal that cannot be found and has never been proven by either the private citizen or the scientific community sounds completely absurd! Couple that with the knowledge that such an animal and similar creatures have been claimed sighted by one and more witnesses right around the world for many years but never brought into captivity adds even less credibility to the idea of its existence.

The Sasquatch, however, is most certainly not the only creature sighted but never confirmed as actual. Cryptozoology is the formal name for the study and research of unproven species. Such animals as the Ogopogo and the Loch Ness Monster as well as the Sasquatch have eluded capture despite ongoing research and upgraded testing methods. It would be easy to dismiss all the above as creatures of the imagination unless it is understood that the science of Cryptozoology is not a well staffed field; there are very few serious researchers looking into claims of creatures not featured in anyone's zoo.

While attending the 1989 conference on Cryptozoology, at the University of (Pullman) Washington I had the opportunity to meet a man extremely well known for his research into the Sasquatch. Rene Dahinden has been looking into reports and following up sightings for nearly thirty years. He co-authored the book "Sasquatch" (1973, Dale Hunter, Rene Dahinden) and has been credited with being the first person to devote himself completely to the finding of the Sasquatch. To this date Mr. Dahinden has not had the experience of actually seeing one.

NORTHWEST TERRITORIES
ALBERTA
FORT VERMILLION
BRITISH COLUMBIA
GRANDE PRAIRIE
VALLEYVIEW
ATHABASCA
DRAYTON VALLEY
VEGREVILLE
JASPER
EDMONTON
SASKATCHEWAN
RED DEER
DRUMHELLER
CALGARY
HIGH RIVER
MEDICINE HAT
LETHBRIDGE
TABER
SIGHTINGS
FOOTPRINTS
N
UNITED STATES OF AMERICA

With very few people working on the discovery of a Sasquatch or the confirmation of its non-existence, the fact that even common animals often elude sighting for long periods of time can add to the possibility that the right person has just not been in the right place at the proper time. To give an example of what I mean by common animals eluding sighting I will briefly take you back to the spring of 1986.

In 1986 I was serving with the army First Battalion P.P.C.L.I. and stationed in Calgary. On the day of our practise for the Regimental Ceremony The Trooping of the Colors, a small private plane crashed near a mountain in the Kananaskis region.

The Kananaskis area is a very widely used recreational park housing open fields, small towns, the foothills of the Rocky Mountains and miles of hills, dense trees and open fields. Tourists from around the globe visit Kananaskis on a regular basis and locals use it as a get-away from the whirl of city life.

When the plane was reported to have crashed, our government ordered the Air Force, the Army and the R.C.M.P. to begin a search of the area it was to have gone down in. We set up camp in Sibbald Flats campground and proceeded with a ground search. During the search 150 troops covered an area of 900 square miles of mountains, valleys and bush. The R.C.M.P., and the Air Force covered the area by helicopter and plane and several civilian volunteers also combed the ground. Hikers, campers, mountain climbers and a crew of people preparing Mount Allan for the 1988 Winter Olympics were also within the Kananaskis area. The plane, its pilot and passenger were not found until just before the search was to be abandoned.

During the search, the group I was with saw only one bear and it was from a considerable distance. Yet, we had covered an incredibly large territory known widely for its bear population!

Taking into consideration the vast number of people and equipment nearly unsuccessful in finding an airplane, its pilot and passenger in a Provincial Park - would it therefore be unreasonable to suggest that with only a few people looking and an area that is, essentially, the entire global surface of the earth, the Sasquatch search is not entirely unsuccessful? The Sasquatch search has yielded some results. Witnesses world-wide have sighted the creature, two men have filmed the creature and footprints have been found, photographed, reported and cast into molds.

Professor Grover Krantz of Washington State University estimated, as said earlier, some years ago, that there may be as few as 200 Sasquatches living in the Pacific Northwest of the United States. Professor Krantz is one of the few scientists who are convinced that there is something to the story of the Sasquatch.

Other members of the scientific community who are skeptical of its existence have stated that they cannot research something that has no claim to fact. Unfortunately, these scientists do have a point. If science devoted itself to proving the claims of a few unconfirmed reports of creatures not previously known, Vampires would be a major field of study.

On the other hand, the Sasquatch has been reported as far back as the discovery of North America. It was then that the Indians told the rest of the world about their creatures.

David Thompson, who was believed to be the first white man to find the tracks of a Sasquatch, kept a narrative (diary) which documented his encounter in 1811. One entry, dated January 7th was written about tracks he found in the snow while crossing the Rocky Mountains through the Yellowhead Pass near the present sight of Jasper. At the time, he was travelling with other early explorers as well as some Indians. A part of the narrative is as follows:

"Continuing our journey in the afternoon we came on the track of a large animal in the snow, about six inches deep on the ice. I measured it: four large toes each of four inches in length. To each a short claw; the ball of the foot stuck three inches lower than the toes. The hinder part of the foot did not mark well, the length of fourteen inches by eight inches in breadth, walking north to south and having passed about six hours. We were in no humor to

Author searching forestry road, Banff National Park.

follow him. The men and Indians would have it to be a young mammoth and I held it to be the track of a large grizzled bear, yet the shortness of the nails, the ball of the foot, and its great size was not that of a bear, otherwise that of a very old bear, his claws worn away. This the Indians would not allow." (Thompson, David p. 36)

Strangely, the Indians with him did not report the track to be that of a Sasquatch. This has led some to believe that it indeed was not a track of the creature. The size of the print, however, could not be easily identified with any other animal. The prints were reported to be 14 x 8 inches in size - not the shoe size of your average grizzly bear!

It would be safe to assume that a group of explorers passing through the Rocky Mountains would have encountered hundreds of animal tracks as well as many animals on their journey. Accepting the fact that there were no automobiles in 1811 and travel was done on foot or horseback, it would also be safe to assume that David Thompson and his men were travelling a vast animal-laden area for an incredibly long time (compared to modern travel standards). For Mr. Thompson to note the finding of tracks as an unusual occurrence he would either have to have lost his marbles, been exceedingly bored, or discovered a set of tracks with such rare qualities that they demanded both his attention and the time it would take him to note them in his narrative.

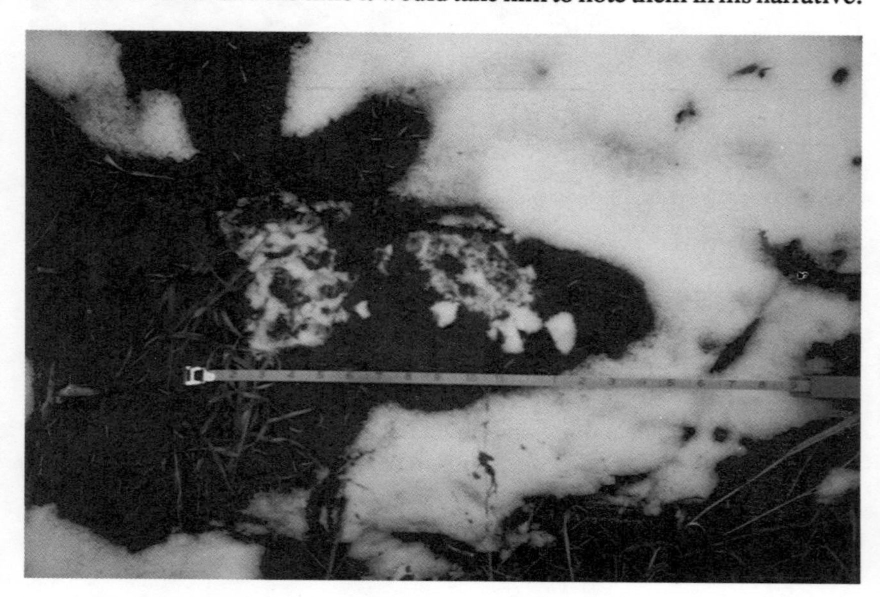

Footprint found around Baptiste River (northwest of Rocky Mountain House, Alberta) (October 1989).

David Thompson not only wrote of the tracks once, but on his return trip the following autumn he mentioned them again.

"I now recur to what I have already noticed in the early part of last winter, when proceeding up the Athabaska River to cross the mountains, in company with ... the men and four hunters, on one of the channels of the river we came to the track of a large animal which measured breadth by tape line. As the snow was about six inches in depth, the track was well defined and we could see it for a full one hundred yards from us. This animal was proceeding from north to south. We did not attempt to follow it, we had not the time for it and the hunters, eager as they are to follow and shoot every animal, made no attempt to follow this beast, for what could the balls of our fowling guns do against such an animal. Report from old times had made the head branches of this river, and the mountains in the vicinity the abode of one, or more, very large animals, to which I never appeared to give credence; for these reports appeared to arise from the fondness for the marvellous so common to mankind; but the sight of the track of that large beast staggered me, and I often thought of it, yet never could bring myself to believe such an animal existed, but thought it might be the track of some monster bear."

Although David Thompson's narrative is difficult to read and his grammar less than adequate by today's standards, the state of mind and his belief in what it was is impossible to ignore. Thompson believed that he had

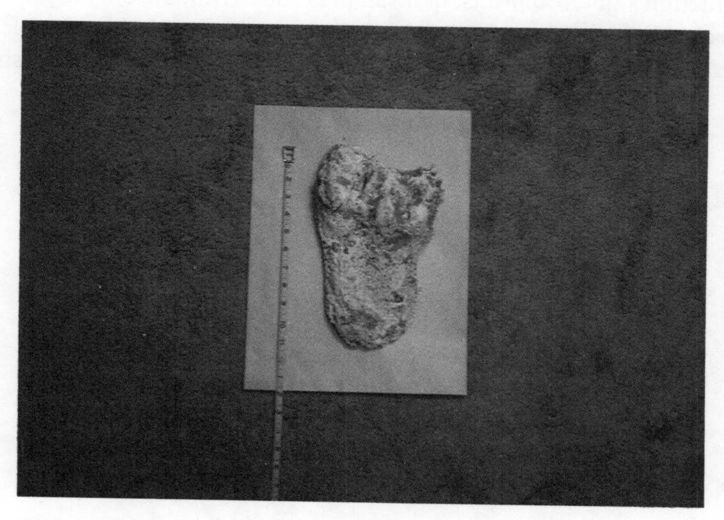

Casting of footprint found along Baptiste River, October 1989.

encountered the tracks of the beast which was commonly told of by the Indian peoples. According to Thompson, the tracks were large enough and deep enough to scare even his blood thirsty hunting party off from pursuit.

Thompson measured the size of the tracks with a tape measure, he did not guess at the size; then he recorded the shape, depth and on going length of the prints in the snow. Observing that these prints could be seen from 100 yards and were embedded in what he refers to as "ice" the animal who made those tracks must have been extremely heavy and incredibly large.

If Thompson and his men found the tracks of a bear the bear broke all records for size and set a new time record for a bear walking on its hind legs. Neither the hunters nor the Indians travelling in his party could adequately explain to Thompson's satisfaction what the species of animal was. Thompson had heard the stories, however, and by autumn had concluded that what he saw was the very same animal we now call the Sasquatch.

With David Thompson's report, the witnesses of the Big Horn Dam construction sight, footprints, hair samples and a film the only mystery left uncovered is why science has not yet begun a serious study into the question of the Sasquatch to determine, once and for all if such an animal exists.

CHAPTER THREE
"More Than A Legend"

In August of 1988, the Warden's office in Waterton Lakes National Park received a report from Darwin J. Gilles claiming that he, his girlfriend, along with a friend and his wife saw an unidentified animal the night before. The report read as follows:

> "At approximately 12:50 a.m. at the Crandell Lake Campground we spotted a very unusual animal. We were sitting at our campfire when we heard some snorting. We assumed it was a deer, but upon further observation we decided it was a bear and bolted for the cars. The animal was on its hind legs and we switched on the headlights on one of the vehicles. From the shadows I could see the animal was moving on its hind legs so I called to the other vehicle to turn on their lights.
>
> What we saw then was incredible. This animal was not only on its hind legs, it was striding (like a human) we watched as it walked through the trees for at least 3 to 4 seconds.
>
> I immediately thought it was a joke. We're all convinced it was not a bear. We jumped into the same vehicle and followed in (its) general direction...
>
> We came across another vehicle and flashed our lights. These people had also sighted something very strange and were quite scared. This confirmed that we had all seen something.
>
> It is important to note that we are four mature, responsible and professional people. We thought very carefully before coming in to report this incident at the warden's office. All four of us are convinced that it was not a bear. I am equally convinced it was

not a practical joke. If it was, it was pretty elaborate and well done.

From our sighting, the best description we can give is as follows: the animal was approximately 8 feet tall (as measured by the tree it was standing beside in our campsite). The animal was never on all fours. When we switched on the headlights and got a good look, this thing was striding... It also had long arms which were swinging while it moved through the bush. It wasn't a bear, okay.

I don't know what more I can write about this incident. We would appreciate hearing anything that might explain what we saw (or additional sightings, if any).

Darwin Gilles"

On the night of August 29th, 1988, Mrs. Susan Stoness, the first member of the Gilles party to sight the creature, responded to my newspaper ad for information on the Sasquatch.

Susan told of the incident, which involved her, her husband Scott Stoness, Darwin Gilles and Shannon Senkow, while on a camping trip in the Crandell Lake campground located in Waterton Lakes National Park in the southwest corner of Alberta. We proceeded with the interview.

On the evening of the sighting the two couples were attempting to play a game of hearts (cards) on a picnic table next to the fire pit by their campsite. The wind picked up, however, and blew the cards off the table scattering them across the ground. At approximately 12:50 a.m. they gave up the game and decided to turn in for the night. Susan and Scott headed down a path towards the public washrooms to brush their teeth.

Within moments Susan was frightened by a noise. Scott, knowing his wife was afraid of being in the wilderness at night, paid no attention to her complaint, took her by the hand and continued down the path. Then, at the same time they both saw a strange creature standing on the trail ten feet ahead. The animal made a low grunting noise. Susan yelled "It's a bear", tore away from Scott, ran back down the path toward their campsite. Darwin and

GORILLA FOOT

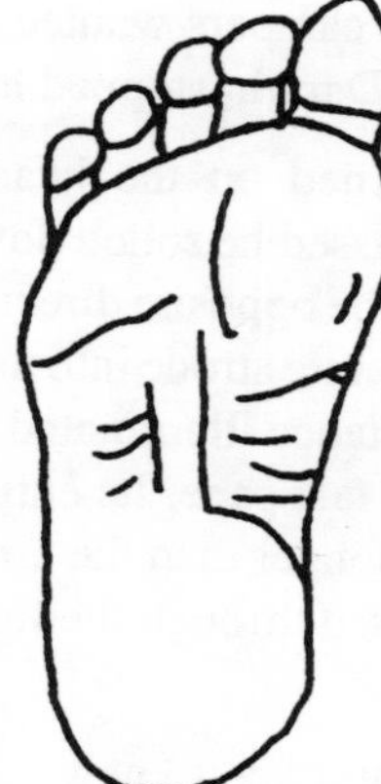

SASQUATCH FOOT

HUMAN FOOT

Shannon, who had not yet left the firepit heard Susan yell, stood up and turned around in time to see her flying back down the path toward them.

Panic struck. All three headed for the two cars. Shannon fled to one car while Darwin and Susan headed for the other vehicle. Meanwhile, Scott was still on the path, ever so slowly backing away from the creature. The creature moved off the trail and into the trees. Scott made it back to the campsite and ran for a car. He ended up in the same vehicle as Shannon - his wife would not open the car door to let him in.

Darwin sat in the car with Susan trying to keep her calm. Susan, shaking with fright, worried about other campers wanted to "lay on the car horn to wake and warn the neighbors". Darwin stopped her.

After awhile Darwin turned on the headlights. He saw nothing. Believing that the danger had passed he rolled down the window and yelled for Scott, whose car was facing the opposite direction. Just then an eight foot high thin hairy man-ape-like creature strode into the area lit by the headlights. In four strides it covered the distance illuminated by the lights, not running, just walking ... at an incredibly fast pace. Its entire body was covered with black hair. Its arms were much longer than the arms of a human. In a matter of seconds the creature had passed through the light and back into the trees, not to be seen again.

It was then that Darwin and Scott knew what it was they had all just witnessed - the legendary, apparently fiction animal known as the Sasquatch. The two men wanted to get out and search for tracks but Susan and Shannon were too shaken, they just wanted to pack up and get out. After some insisting, however, a compromise was made and the four campers drove around the area trying to spot the creature again.

In their effort to get another glimpse of the creature they came across a truck load of people who reported having seen "something strange" on the grounds an hour previously. The people in the truck did not give their names and no further reports were heard about the incident from these people.

After a night of restlessness and occasional sleep the Gilles party contacted the warden's office who followed them back to the area of the sighting to search for tracks. The warden kept asking if there was a possibility that the creature had really been a bear. Not one of the four witnesses were willing to concede to a bear story. In their words "It didn't look like a bear, it didn't walk like a bear, it didn't even resemble a bear". There was just no evidence to substantiate the creature being a bear.

Susan coerced Darwin and Shannon to take part in the interview at her and Scott's home. Not surprisingly, the four witnesses had more questions to ask of me then I did of them. Each person was interviewed alone in a separate room and asked for their version of the sighting.

Following are excerpts of the interviews conducted:

Susan Stoness:

Q: Describe what you saw.

A: (I saw) a creature first from about 10 feet away in the outskirts of the campsite, located approximately 15 miles outside of Waterton. My first reaction was that it was a bear. He grunted as we approached. He was standing on two legs and covered with black or dark brown hair. I would estimate the height to be eight feet tall, (in accordance to the height of the tree we later measured against).

He was heavy, huge, maybe six to eight hundred pounds. His face was flat with eyes, nose and a mouth. The arms were very long and the tips of his fingers came very close to the knees.

Q: How long, in total did you see the creature?

A: Approximately one or two minutes.

Q: When you saw the creature did it react in any way, and if so, exactly how?

A: He grunted at us ... Not seeming hostile or scared. (I think) he was letting us know that he was there, almost like a warning.

Q: In your own words describe what happened.

A: *...My husband held me by the hand and as we started down the trail I thought I heard something and I told him, "I think I hear something". He really didn't take notice of that because I always hear stuff; I'm kind of scared at night.*

We took a couple more steps and then we saw a big hairy creature standing up in front of us, probably ten feet away. Then it grunted at us three times. Well, I screamed, "It's a bear," and ran to the car. Darwin got in the car with me, while Shannon ran into the other car. Scott, when he finally got to us, tried to get in the car with us but I wouldn't open the door so he ran to the other vehicle and jumped in with Shannon.

... The two cars were bumper to bumper and when we put the headlights on and looked around Darwin thought he saw something moving by the fire towards the trail. We yelled for Scott to turn on his headlights. About 10 seconds later the creature walked into the light. It was walking on a ridge about 30 to 35 feet in front of the headlights. It looked back at us but did not break it's stride. It wasn't really fat like some of the pictures we have seen, in fact it was slender with really long legs, disproportionate to the body. My husband yelled "holy ... that's incredible, it's a Sasquatch".

We drove around the campground and came to another truck full of people flashing their lights at us. The four people in the cab of the truck listened to our story and claimed to have seen something similar about 20 minutes earlier, but they were quite drunk and only three of the four actually saw it. When we got back to our campsite I did not want to get out of the car or stay the night in that place. Scott got into the car with me and fell asleep. I stayed up all night and watched the trees.

Darwin and Shannon stayed up all night too by the fire. The next morning, after much debate, we decided to report it to the warden. We went to his office and waited for him, told our story and Darwin gave the report. The warden tried to convince us that we had seen a bear. He kept saying "well, bears stand up on their hind legs, are you sure it wasn't a bear?" He was nice, but I think he thought we had seen a bear.

Scott Stoness

Q. When and where did this sighting take place?

A. *The incident occurred in Waterton Park, Crandell Lake Campground. It was the May long weekend, early (before 1:00 a.m.) on Monday.*

Q. Could you describe the creature?

A. *It was walking on two legs, I never saw it go down on all fours. Its hair was dark brown or black and it blended into the darkness - I didn't actually see it until we got very close. The creature was between 7 1/2 and 8 feet tall and approximately 500 pounds.*

The second time I saw it, while in the car, I noticed its stride was about 5 feet. It moved very quickly through the beam of the headlights, keeping its legs almost straight.

Q. Your wife mentioned that it made a noise. Could you describe the noise?

A. *The sound it made reminded me of that of a bull when he is in pursuit. I was chased by a bull once and the sound it made is the closest I can come to describing the creature's noise. I do not think that any human could have made such a sound - it was like an animal with a big throat blowing a lot of air.*

Q. Once Darwin filed the report, the warden returned with you to look for tracks. What was found?

A. *Well, the area was very rocky and covered in moss. I think an elephant could have trampled through unnoticed. We did not find any tracks to substantiate the sighting, but considering the terrain I am not surprised.*

Q. At what point did you believe this creature was a Sasquatch?

A. *The second time I saw it I was pretty excited because I thought "this must be what everybody says is a Sasquatch. Both Darwin and I wanted to chase after it and see what it really was but my wife wasn't too keen on the idea. Instead we drove around awhile. A big truck with 3 or 4 people in the front approached flashing their lights at us. They asked if we had seen anything. So we described our creature and they described theirs. They also thought the creature was a Sasquatch. I don't think they saw it as well as we did though. When we told the warden he tried to convince us that our creature was a bear.*

Darwin Gilles:

Q. When and where did this incident occur?

A. *On May 23 at 12:50 a.m., site C-3, Crandell Lake Campground, Waterton National Park. It was a typical campground with a circular gravel road. We were on an elevated site about three feet up from the main path.*

Q How far were you from the creature?

A. *When I first saw it I was roughly 20 to 30 yards away.*

Q. What do you think this animal was doing in the campsite?

A. *Initially, when Scott and Susan left the campground walking towards the creature it grunted. They thought it was a deer and continued walking, then it grunted again. I believe that the creature was just curious and I'm convinced that it was watching us for quite awhile. The reason being is when the cards blew off the table about 15 minutes before Scott and Susan saw the animal, I remember hearing something. I am convinced that it was just curiously watching us.*

Q. From what you remember, describe what happened that night.

A. *We were playing cards by the campfire and Scott and Susan decided to turn it in. They started off down the path to brush their teeth. As soon as they walked out of the campsite we heard the first grunt and I jumped up. Scott yelled back "it's just a deer" and grabbed Susan by the hand continuing down the path. Then this thing grunted again. This time*

Susan said something like "that's not a deer, its a bear!". As soon as Shannon and I heard that we were up. Shannon suggested we move calmly towards the cars but it was like a free for all, we sprinted. In the confusion, I ended up in the same car as Susan; Scott ended up in the other car with Shannon. Our car was facing inward toward the bush and away from the direction of the first sighting. Susan was quite excited and wanted to honk the horn but I stopped her and turned to look back in the direction of the bear. Then I saw a tall skinny shadow. "What is a bear doing on its hind legs" I thought, "usually they come down off their hinds if they are standing up". I turned on the headlights but couldn't see anything, so I rolled down the window and yelled for Scott to turn on his. The four of us saw this thing plain as day. It was leaving our campsite walking on an angle away from us across the beam of the headlights and into the trees. There is no mistaking what we saw.

Shannon Senkow:

Q. Can you remember what your initial reaction to the sighting of this creature was?

A. *Yes. I thought it was a bear so I was scared and ran for the car.*

Q. When you finally did see this alleged bear, what did it look like?

A. *It was standing there covered in dark hair. It had very long and hairy arms and stood a good 7 or 8 feet tall. With the headlights of the cars on I saw that it wasn't a bear. It took about ten steps in a human-like gait and went on its way.*

Four people saw something in the middle of the night far away from the city they lived in. Is it possible they could have seen a bear? Taking into consideration these four people are professional, intelligent careerists I personally believe that it is ludicrous to assume that they do not know what a bear looks like!

The only possibility of these people having not encountered a Sasquatch is that they were victims of a clever and well prepared practical joke. The four people in a truck could have played such a prank and returned to the scene to see if their theatricals were effective. This in my opinion is the only other possibility and giving consideration to the size, stride and noise of the creature, the practical joke theory does not hold much weight.

The life size statue of a Sasquatch at the Museum of Natural History in Banff, Alberta. The general body proportions seem to be correct, however, I disagree with the skin colour on the face and hands.

At the time of writing, I am trying to contact the warden who received this report and went looking for tracks in order to verify that no further evidence was found.

The "Legend of the Sasquatch" which began as an Indian Legend and mountain region folklore has evolved into a serious mystery which is gaining in credibility by the day. In the eyes and ears of intellectuals it is one thing to hear of distant tales told by distant people but it is yet another matter to have these tales substantiated by other intellectuals who appear to have no personal gain in adding fact to what was once mere fantasy.

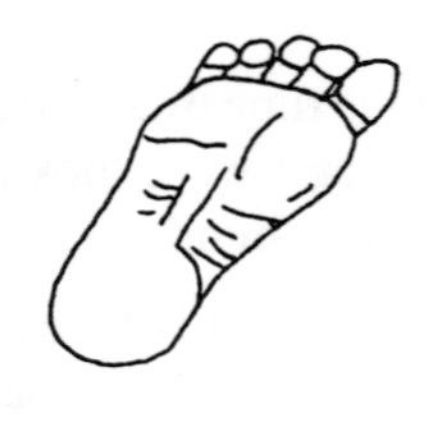

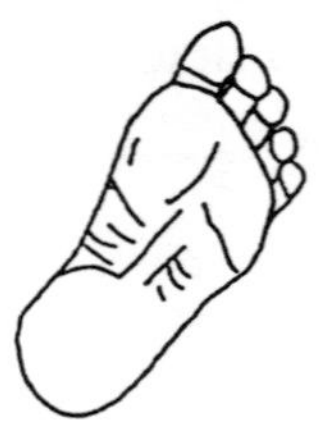

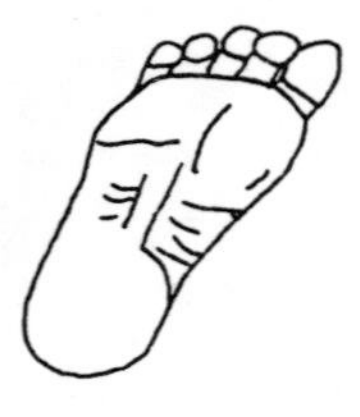

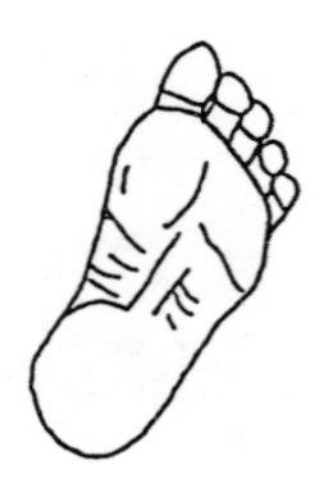

CHAPTER FOUR
"The Tallest Man On The Mountain?"

As reports of the Sasquatch become more and more frequent the question of height raises more and more concern. Reported sightings of the creature are almost contradictory in this area, giving the animal's height from a mere 7 feet tall to a whopping 15 feet tall. Both those who have made height surveys and those who guess verify that an extreme difference in height exists.

Investigators of the Sasquatch most often assume the animal to be between 8 and 9 feet in height. In the past, reports that placed its height in the area of 12 and 13 feet tall were regarded as excited over estimates made by those who were too surprised to give an accurate record.

From Northern California to Washington State and up into British Columbia, the most frequently reported height is between 7 and 9 feet. Many Alberta reports also fall into this height category. In Alberta, however, an increasing amount of sightings claim that the creature stands upward of 12 feet. Nearly one quarter of all Alberta sightings have placed the creature into an unusually high height category.

One sighting claims that the entire head of the animal could be seen above young 9 foot trees. The Big Horn Dam incident placed the height at 15 feet. The young man cutting wood with a band of Indians led by Chief Joe Smallboys saw a 12 foot tall creature at a distance of 300 yards. The two creatures sighted by a man on Highway 11 at a distance under 200 feet were reported to both stand over 12 feet.

Do these contradictory sizes take away from the possible authenticity of reports, or could it be that the Sasquatch is no more unique than the human man in this area? The average height for a man is 5'11" - yet we have a large population of men over 6'5"! The NBA boasts of players over 7' tall. Could

**A six foot two inch man,
in comparison with an 8 foot Sasquatch.**

it be possible that the Sasquatch reaches the height of 8 to 9 feet tall and the female of the species grows to a height of just over 7 feet - with the odd individual reaching heights of 12 feet and over? I personally believe that this is quite feasible. But why are so many of them seen in Alberta? Of this I could only guess - and guess I think I will... A possible reason for the Sasquatch reaching such unusual heights in Alberta could be a mere coincidence.

It is true that in British Columbia reports of 12 foot and over Sasquatches are quite rare. It is also true that the terrain and atmospheric conditions are quite similar across the border of Alberta and B.C. However, B.C. reports far out number Alberta reports and by coincidence Alberta sightings could have just been, thus far, of taller animals than those of our neighbors. It is also important to take into consideration that the excitement of the moment could cause inaccurate guess work on both sides of the border.

There is, however, very little known about the Sasquatch at this time. Until the animal is discovered and researched the question of height will remain a mystery. Possible reasons for the differences can only be guessed at in light of what man now understands of the adaptability of animals in general.

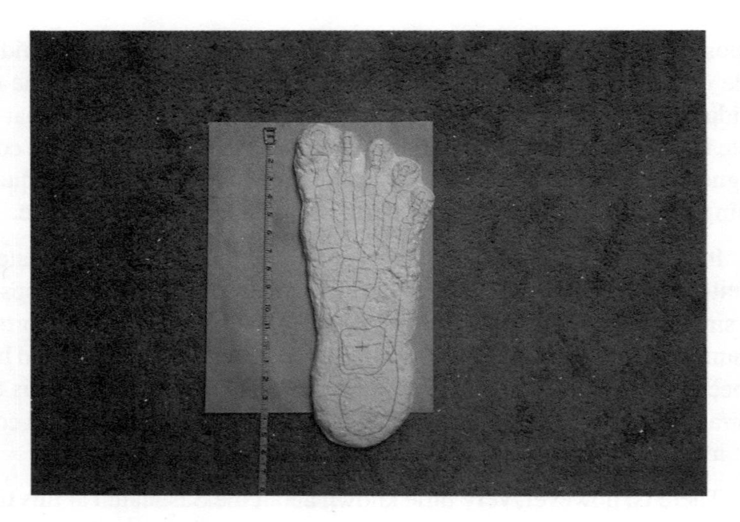

Cast of Bossburg Cripple (left foot). Found in northeast Washington State, 1969/70.

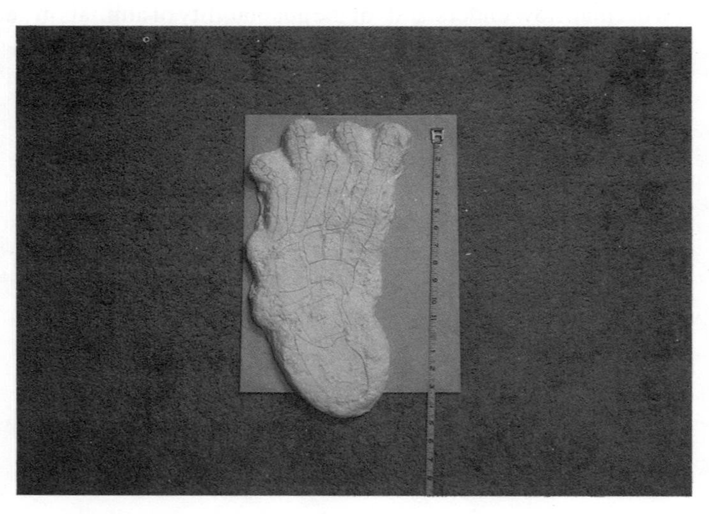

Cast of Bossburg Cripple (right foot). Found in northeast Washington State, 1969/70.

CHAPTER FIVE
"Should There Be A Doubt.."

In the years that the Sasquatch animal has been pursued, it has become evident that its natural habitat is among the trees, in the mountainous regions and away from civilized man. Still, there has been reported sightings of the creature wandering onto private property near and just beyond the cities, ignoring thus far observed caution and coming dangerously close to civilization.

Reports of the Sasquatch peering into windows, approaching campsites, even opening the flaps of a tent left unzipped are on file. Incredibly, there are incidents known of where the animal has run beside a moving car (possibly to get a closer look at the driver). Some have been known to shake vans, trucks and campers. Owners of these vehicles have heard noises at night and found large human-like footprints encircling their vehicles the next morning. Is it possible that the Sasquatch is a Curious-George who stalks the night searching and exploring?

On the night of November 26, 1987, at approximately 8:30 p.m. a 14 year old girl phoned me to report a sighting which involved her and a friend, that took place outside of Airdrie Alberta the previous evening. My own Curious-George nature got the best of me as Airdrie is mainly flat prairie land sparse in trees and even lacking enough brush to hide a jack-rabbit for any length of time.

The girls, who did not wish to be identified in publication, claimed to have seen a Sasquatch wandering around. For the purpose of clarity I will identify the caller as Jill and her friend as Ann.

The two girls had climbed a high fence on the extreme western edge of town that overlooked the railroad tracks. On the other side of the tracks was a smaller fence about 3 feet high. Beyond the fences the girls had a clear view of farmers fields and the distant foothills and mountains. Both Jill and Ann were forbidden by their parents to cross over the fence, and in the spirit of youth, they crossed over it anyhow. Ann climbed over the larger fence, crossed the tracks, climbed an embankment and started for the smaller fence. When Ann had reached the second fence, Jill started her climb. Halfway up she heard Ann screaming and saw her flying back over the tracks yelling "Let's get outta here!"

Both girls scrambled back over the fence and ran into town. Jill confused by the panic finally stopped Ann and asked for an explanation. After Ann calmed down she shared a story with her friend that terrified and excited both of them.

As Ann was about to climb the smaller fence she heard a noise in the farmer's field ahead. Approximately 100 feet away a large black man-shaped creature stood up from a crouch to an incredible height His whole body was covered in hair and he was taller and wider than any man she had ever seen.

Jill, still climbing the first fence had seen nothing but knew by the degree of Ann's fear that her friend had seen something.

The next day Ann was still in a state of fear. Later that day when Jill phoned me to report what had happened I could hear Ann in the background. She really didn't want to talk to me and was not too pleased with the fact that Jill did.

Myself and two others paid a visit to the site the next day. We could see the foothills in the distance and estimated them to be approximately a half a days walk. The ground was still frozen though, and we found no tracks or evidence to back the girls story.

Common sense dictates that a story like this should be viewed as a possible prank. According to Jill no one else knew they were out there so it is not reasonable to assume that a third party played a prank on the girls. Instead, a sensible question would be "Are the girls playing a prank on Sasquatch investigators?"

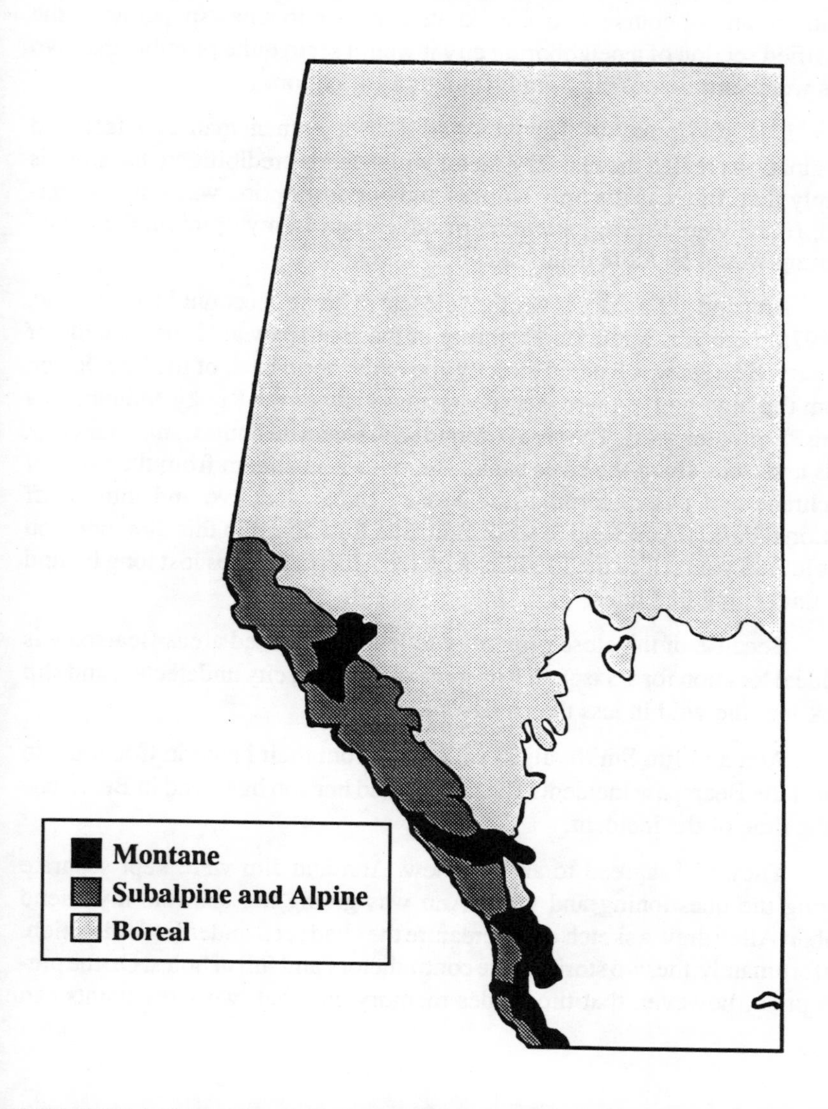
Forest Regions of Alberta
Montane
Subalpine and Alpine
Boreal

Taking into account their age, the fact that they did not want to be identified and of course their immediate response to a newspaper ad in the classified section of a neighboring city it would seem quite possible that two girls were simply looking for adventure and attention.

It is also possible that Ann had seen a human man and fear and imagination created the rest. The factor which lends credibility to the story is simply that Ann was the only witness and her description was not exaggerated. If she were seeking attention or adventure, in my opinion, the actual sighting would be more detailed.

Ann and Jill's story is not the "closest to home" account I have on file. In 1978 there was a reported sighting in the Bearspaw area just outside of Calgary. Bearspaw is located directly above the north bank of the Bow River. From the high bluffs, there is a spectacular view of the Rocky Mountains a mere 35 minutes away. The area is dotted with beautiful homes, small ranches, hills and trees along the river bank. The river is upstream from the town of Cochrane and passes through the Stoney Indian Reserve and into Banff National Park. If you were to walk along the Bow River in this direction you would find yourself in the foothills with the Alberta Prairies lost long behind the thickening tree line.

Because of the close proximity to heavily wooded areas Bearspaw is an ideal location for a Sasquatch to come close to the city undetected, and slip back into the wild in less than a day's journey.

Ann and Jim Smith (alias) called me from their home in Cochrane to report the Bearspaw incident. The mother and her son had lived in Bearspaw at the time of the incident.

They both agreed to an interview. Ann and Jim were kept separate during the questioning and while Ann was giving her answers my friend Robert Alley drew a sketch of the creature they had seen under Jim's direction. Unfortunately, the two stories were contradictory and full of holes. On the presumption, however, that time fades memory and that two view points can

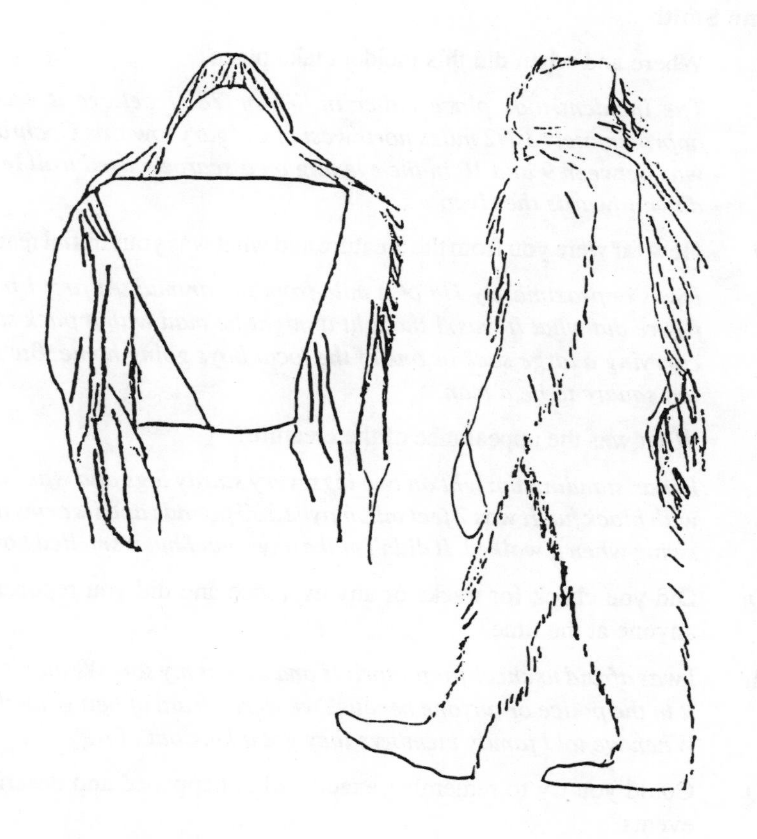

Sketch of the creature seen by Jim Smith, drawn by Robert Alley under Jim's direction.

create confusion, I have included excerpts of the interviews to allow the reader to decide for himself.

Ann Smith

Q. Where and when did this incident take place?

A. *The incident took place either in '77 or 78 (I believe it was 78), approximately 3 1/2 miles northwest of Calgary towards Cochrane. It was between 9 and 10 in the evening on a scarcely used trail leading down towards the river.*

Q. How far were you from the creature and what was your initial reaction?

A. *I was approximately 1/8 of a mile from the animal. At first I tried to figure out what it was. I thought it might be man with a pack sack or carrying a large sack or one of the local boys going home. But it was too square to be a man.*

Q What was the appearance of the creature?

A. *It was standing upright on two legs, very sturdy legs and was covered with black fur. It was 7 feet tall, maybe 300 pounds and its arms did not swing when it walked. It didn't make any sound but it smelled horrible.*

Q. Did you check for tracks or any evidence and did you report this to anyone at the time?

A. *I was afraid to check for footprints and so was my son. We didn't report it to the police or anyone because we were afraid of being laughed at. When we told family members they got a kick out of it.*

Q. Could you try to remember exactly what happened and describe the events?

A. *The dog started to whimper outside so I figured that some other dogs were picking on him. I went outside and he was under the porch whimpering in a way that I never heard him whimper before. Then the odor struck me. It was awful. I called my son and we got the dog out and continued with what we were doing. When I went over to the french windows I noticed that it was snowing and then I saw this thing walking in the direction of the river. That would take him across the Bearspaw road like in that area down the Transalta Incemanation Station. It was walking very quickly. I called to my son and said "Could you see a packsack on his back?" because it looked so very big. My son said it looked like it was covered in black animal fur. Its legs were very stocky, it took big strides, it was very squared in the upper body area and it just kept walking until it was out of sight.*

Ann's story sounded quite authentic and so did Jim's, only their stories sounded like two different sightings! When I had completed the interview with Ann I called Jim into the room and this is the version he gave...

Jim Smith

Q. When and where did this incident occur?

A. *November, 1978 in Bearspaw around 10:00 p.m. The area was hilly with flat grass and stubble grass. I was about 200 yards from the creature.*

Q. Could you please describe your initial reaction and what the creature looked like to you?

A. *My first reaction was that it was BIG, real BIG! It was walking on two legs with the upper part of the torso hunched. The creature appeared to be covered in black hair, standing approximately 6 to 7 feet in height and about 280 pounds Its arms hung to its side and there was a smell like sulfur.*

Q. Did you check for footprints or report the incident?

A. *No and No.*

Q. Would you describe exactly what took place.

A. *At about 10:00 p.m. my mother and I were sitting in the house and the dog started to bark. I went outside to see that the problem was. He was a guard dog so he only barks when something is in the yard. I went out to the porch to calm him down and dog shot into the house whimpering. I started to smell this stench and I looked up across the road. I saw this huge figure and knew it was too big to be a person. I went back into the house and onto the front porch and we watched it walk in a southward direction toward the river.*

Q. Did you see it first or your mother?

A. I did.

Q. Then you told her and she came out to watch it with you?

A. Yes.

Both Ann and Jim claim to have seen the creature first. Ann through the french doors and Jim on the porch. Ann claims the dog was whimpering under the porch and she went outside to see what was the matter with it. Jim's story makes no mention of the dog being under the porch or Ann going out to see it. Jim described the upper torso as being hunched, Ann saw it standing very straight. Ann saw it with Jim for 4 to 5 minutes while Jim saw it with Ann for about 2 minutes.

Either one or both of the alleged witnesses has forgotten the details of the event or there was no event. They both did agree on the sketch and the details. Their given reason for withholding their true identities is that they claimed to have taken enough ridicule from the family already.

I cannot confirm that this was an actual sighting of the Sasquatch and I cannot prove that it wasn't. If, however, it was the fact that they both agreed that the sketch was the creature in question, a sketch of a Sasquatch, and I know that the Calgary Zoo did not lose one of its gorillas that day, it would leave out the possibility of any other known animal. So, it is up to the reader to decide if this incident warrants the credibility of an authentic sighting, a simple case of fiction or that of mistaken identity. And, despite the time between the event and the interview (9 years) the discrepancies cannot be ignored.

CHAPTER SIX
"Sasquatch Drinks Coors"

As an advertised investigator and researcher into the strange and often humorous subject of the Sasquatch I have received more than a shareful of prank calls, jokes and just plain rude and nasty reports. The jokes and prank calls often lend a side of humor on a subject with very serious applications and implications.

One night I picked up the phone and the caller on the other side identified himself as a having information on the Sasquatch. "When and where did you see the animal?" I asked.

"Oh, I saw him about right now, in fact he's sitting here having a brew with me, want to talk to him?"

Plenty of midnight callers, with nothing more to do than bother others have dialed my number and claimed to see the Sasquatch. Sometimes the stories involve partying with the animal or watching it descend from a UFO. If the animal is a native earthling, the idea of an intergalactic heritage is a little far out into left field (or left galaxy, if you will).

One man (who has, thank heavens, never dialed my number) claims to communicate with the creature using ESP. Other claims made place the Sasquatch out of the realm of reality and into the realm of fantasy. The only story I have yet to hear is that of a Sasquatch riding through the sky on a unicorn.

Those who make these claims and some who go as far as to state them publicly are known among researchers as "the lunatic fringe". Many of this classification have engaged in writing documentaries and claiming authority or sole knowledge of the animal. It is important to note that those in the lunatic fringe have little to no evidence to back up their claims to authenticity or authority. If the Sasquatch is indeed a creature or animal living in some areas of the world, no one man could have complete authority on the subject unless he alone had all the evidence and a Sasquatch to back it up. The subject of this animal is yet on the threshhold of science and will remain there until a body with bones, sinew, tissue and muscle is brought to the inner circle of science and revealed to the outer circle of civilization.

This is one example of how the Sasquatch has become a local tourist attraction in some places. The sign reads "Big Foot Campgrounds", and is located at Harrison Hot Springs, B.C.

Even the B.C. Government has gotten into the act. This Provincial Park is located just outside of Harrison Hot Springs, B.C.

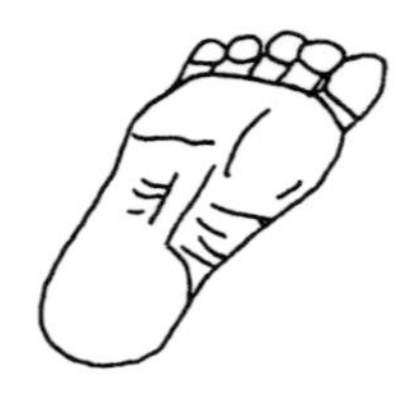

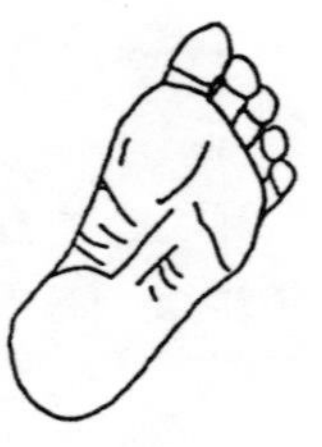

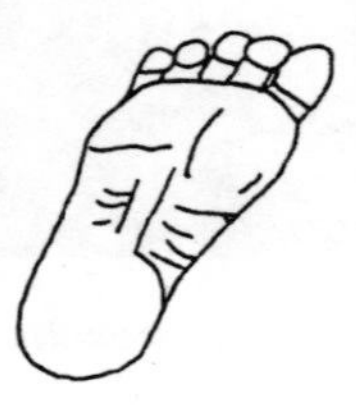

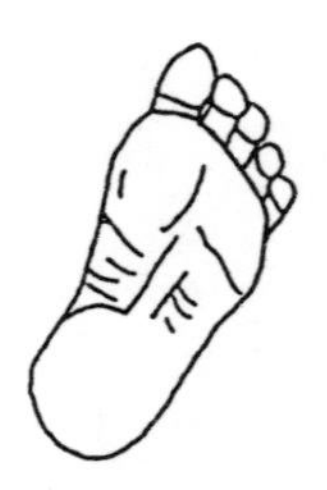

CHAPTER SEVEN
"B.C.s Hairy Giants"

Although this book deals mainly with reports from Alberta, it must be pointed out that the province of British Columbia is the traditional stomping grounds of the Sasquatch in Canada. There have been reports from various other parts of the country. Manitoba and Ontario have had a good many sightings reported over the last 30 years. William M. Borody, Guy Phillips, and Bill Mason (a fellow whom I've not yet had the pleasure of meeting or corresponding with) are well-known for their investigations into reported sightings of Sasquatch-like creatures in the province of Manitoba. It is in the wild mountain wilderness of British Columbia though, where most reports of this creature come from.

The name Sasquatch, is a Coast Salish word which means "wild man of the woods". The native culture of the lower Fraser Valley believed that these creatures were a race of giant people who would come down out of the mountains to steal women and children. The creature had a sort of bogeyman appeal which was useful to make misbehaving children more controllable. You can, today I'm sure, still find the odd elderly Native man or woman in southwestern B.C., who will still remember back to when they were children, being told by their parents or grandparents not to go into certain wooded areas because the Sasquatch lived there.

J. W. Burns, who for years was a school teacher at the Chehalis Indian Reserve on the Harrison River near Harrison Hot Springs, wrote a number of articles in the 1920s and 1930s concerning the Sasquatch. They received wide magazine and newspaper coverage in Canada and the United States. There was one article written by Mr. Burns that appeared in Maclean's Magazine on April 1st, 1929 titled "Introducing B.C.'s Hairy Giants" which resulted in a great deal of public interest and was probably indirectly responsible for the name "Sasquatch" being adopted by the white man in Canada.

Southwestern B.C., the lower Fraser Valley, and the Harrison Lake area, probably have had more reports of Sasquatch-like creatures running around the woods since the turn of the century than all of Alberta put together. Since the early 1980s, sighting reports in this area have decreased to almost zero. Why this is, I have no idea. Despite the lack of recent reports from these areas, there have been numerous reports from elsewhere in the province. The best set of footprints that I myself have seen, were along the Chilliwack River in August of 1986. I was traveling through the lower Fraser Valley at the time, while camping and looking for footprints plus interviewing anyone who would talk to me about things they had seen, or thought they had seen. I was in the town of Hope having a cup of coffee when an elderly gent whom I'd never met before came up to me and asked "Your that Sasquatch nut aren't you?" I nearly spit my coffee all over myself and told him that I would not have quite put it that way, but yeah, I was the Sasquatch nut, I introduced myself and asked for his name. He didn't tell me his name, he just stood there and said "You know, as far as I'm concerned there is no such thing as Sasquatch, but if you really want to waste your time chasing ghosts, you might as well know that one of those things was seen about two or three days ago along the Chilliwack River". I asked him if he could be more specific as to the location and what had been seen. He replied, "Sorry I can't, I was in Chilliwack and that's where I heard the story". I thanked him for the information and offered to buy him a cup of coffee. He just said, "No thanks" and walked out of the coffee shop. That was the last I ever saw of the man. Later that same morning, I drove to Chilliwack to see if I could find out more. First I checked with the R.C.M.P. but they had heard nothing about it. Later, a young woman told me she had heard about the sighting. It was apparently big news by word-of-mouth for a couple of days but had now died down. She told me that it was an American tourist that had seen this animal while he was fishing, however she did not remember where exactly along the river the sighting took place. I probably could have let it drop right there, but

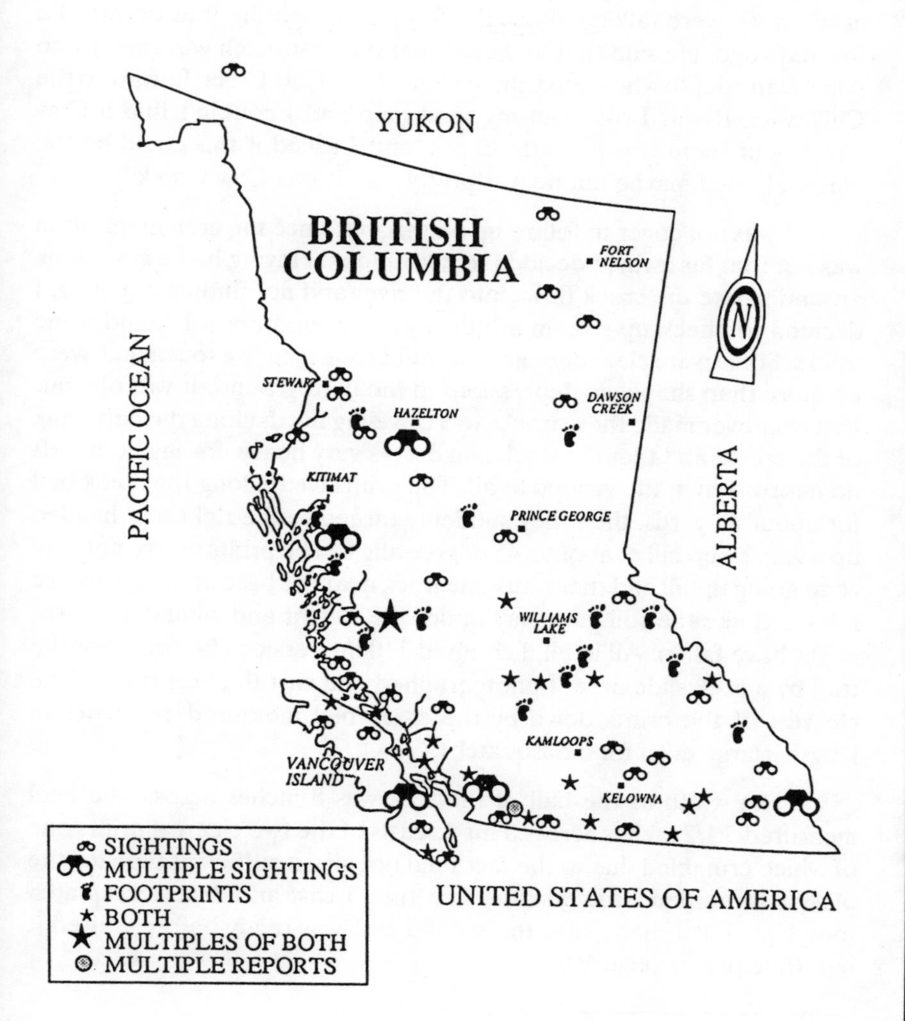
YUKON
BRITISH COLUMBIA
FORT NELSON
PACIFIC OCEAN
STEWART
HAZELTON
DAWSON CREEK
KITIMAT
PRINCE GEORGE
ALBERTA
WILLIAMS LAKE
KAMLOOPS
VANCOUVER ISLAND
KELOWNA
SIGHTINGS
MULTIPLE SIGHTINGS
FOOTPRINTS
BOTH
MULTIPLES OF BOTH
MULTIPLE REPORTS
UNITED STATES OF AMERICA

as we were talking a young man who had overheard us came up and asked if we were talking about the Sasquatch sighting that occurred a few days ago. He said that he heard that the Sasquatch was seen by an American couple who were fishing near where Cow Creek flows into the Chilliwack River. I took out my road map and I couldn't find a Cow Creek, but there was a Cattle Creek, and I asked if this could be the place. He said maybe but he was pretty sure it was Cow Creek.*

I was not eager to follow up on this, but since the area in question was not that far away, I decided to take a look. Having had a good look around where the creek flows into the river and not finding anything, I decided to check up-stream a little ways. Around noon I found some tracks. Some were clear enough you could count the five toes, some were no more than shapeless depressions in the hard ground. It was obvious that whatever made these tracks was traveling north along the east bank of the creek, and that the track maker was very heavy, for my boots left no impression in the ground at all. The prints went along the creek bed for about 40 yards, then they suddenly turned to the right and headed up a very steep hill at about a 45 degree angle. The prints where not very clear going uphill and there was one track near the base of the hill where it looked like the soil gave way under the weight and what ever it was might have fallen. All in all, I counted 110 impressions before I lost the trail by a rock slide area. I photographed and cast the best track. The clearest of the prints down by the creek bed measured 18 inches in length, (large even for a Sasquatch).

The width at the ball of the foot was 8 inches across, the heel measured 7 1/2 inches across. I made casts of the two clearest prints one of which crumbled due to the fact I did not give it sufficient time for the plaster to harden, (this was the first time I cast an alleged Sasquatch footprint) but I made sure the second cast was ready before I lifted it out. (See photo, page 103.)

** Note, I have since discovered that this creek is called Cow Creek on some maps, Cattle Creek on others while some maps don't show it at all.*

Fakery in my opinion cannot be ruled out here. I often wonder if one of the people, especially the lad who told me Cow Creek was the place where the sighting took place had raced out there before me and planted these tracks for me to find. But judging from his thin stature he would have had to carry a lot of extra weight on his back, especially if he was stomping about with carved out wooden feet stuck to the bottom of his boots, and going up hill as well. I don't know if a Sasquatch made the footprints I saw in the summer of '86, however, they did intrigue me enough to keep my interest in this subject going till the present. If I had not seen those tracks, I might have given up on the Sasquatch and gone on to other and - as my ex-wife would put it - more important things. I did learn a lot from that experience, one, is to learn how to use a camera properly. I was disappointed in how the photos turned out. Two, give plaster of paris time to fully dry before you lift the cast out. I still kick myself over that first print. Three, do a much more thorough study of every detail before you leave the area.

In November of 1987, I got a phone call from a gent named Pete Nab, who told me how in the summer of 1973, he and two friends were out hunting in a place known as Hunters Range, about 30 miles east of Enderby, B.C. They were hunting small game with .22 calibre rifles when one of the boys pointed to something about 100 yards away. At first Pete thought it was a bear. He soon realized that this thing was standing upright on two legs. The boys were in the middle of a small meadow, the creature was standing at the edge of the treeline with one hand on a pile of logs. The creature made no threatening moves toward the boys, it just stood there watching them. All three boys, frightened started shooting at the animal. Pete noticed the creature take it's arm away from the log pile and put it's hand to it's chest area as though it was swatting a fly. Other than that, the bullets did not seem to affect it. All three boys then ran in the opposite direction. Pete did look back in time to see the thing turn and walk off the other way. It walked on two legs. When I went to Pete's home to interview him about the sighting, I had the chance to talk to his wife, and she told me that her husband had told this story to other people, but had been ridiculed as a result. He did not strike me as the type of man who made up tall tales. According to some people who knew him, he was the type of fellow who could care less if you believed him or not. I interviewed Pete at his home in Calgary about what it was he saw and at no time did he slip up or contradict himself.

Q. What date did you see this thing?

A. It was high summer, late July, 1973.

Q. It was you and two friends who saw it?

A. Yeah.

Q. Where did it happen?

A. Hunters Range.

Q. That's close to Okanagan Lake?

A. Well, it's about 30 miles from Enderby.

Q. What time of day was it?

A. It was in the afternoon.

Q. Just describe what you saw, and what happened.

A. Well, we were walking through a meadow and we got, I don't know, maybe 25 yards into the meadow. We were looking around, you know with a .22, you hunt little things and we saw, I'm not sure who saw it first, probably all three of us saw it at the same time. It was standing there and it didn't move. At first we thought it was a bear, but it wasn't a bear. We shot at it, it didn't move, it didn't run away from us or attack us or anything. We didn't get any closer to it. We stopped where we were, and we shot at it. We emptied most of our guns at it. Nothing happened, it didn't fall.

Q. No reaction, it just stood there and watched you?

A. Yeah, it just stood there and like it didn't move, it didn't run away or nothing.

Q. Did you three hear it first or see it first?

A. *We saw it. There was no noise. It didn't make any kind of noise.*

Q. You were in the meadow and it was by the treeline?

A. *On the treeline, yes.*

Q. What color was it?

A. *It was black.*

Q. Black?

A. *Yep.*

Q. Was it covered in hair?

A. *Yes.*

Q. Was it standing on two legs?

A. *Yes.*

Q. Could you describe it's face?

A. *The face features that I saw were, you know, thinking back years ago would be, the eyes. They were deep, they weren't, ummmmm, they were deep eyes and the face, ummm. It had a nose, like the nose on it was, ummm, it was a wide nose, like a negroes.*

Q. Like a flat nose?

A. *Yeah, and it had a mouth, and it was big.*

Q. Did you see it's teeth?

A. *No, I didn't see it's teeth.*

Q. What were it's arms like?

A. *They were big arms and they were really long.*

Q. Were they hair covered?

A. *Yes.*

Q. Did you notice the color of it's skin in the facial area?

A. *I didn't see any skin.*

Q. The face was hair covered?

A. *Yeah, that's all I saw was hair.*

Q. As soon as you saw this thing, you started shooting at it?

A. *Yes, we started shooting at it pretty quick. It was only a few seconds, because it was so, umm, we thought it was a bear, but it wasn't a bear, because bears don't lean against trees on their hind legs, and this thing was leaning on some tree logs standing up, and it was bigger than a bear.*

Q. After it was gone did you look for footprints?

A. *No, we didn't. We were just kids then and you know, it scared us because when you have a gun and you shoot something, it's supposed to die. This thing didn't die. After we shot at it, I turned around and strongly suggested that we leave. We all started heading back and we were, well you know, we didn't want to get hurt. I turned around and I saw it turn, it turned quickly, for something that big, it turned fast and walked off into the bush. It wasn't running, but it did move pretty fast. It didn't come after us.*

Q. Did it walk away upright? Did it ever go down on all fours?

A. *Never.*

Q. How tall would you say this creature was?

A. *8 1/2 to 9 1/2 feet.*

Q. 9 1/2 feet tall?

A. *Yes.*

Q. Check for any footprints? (Second time I asked him this.)

A. *No, we didn't. We were scared, eh.*

Q. Did you smell anything?

A. *No.*

Q. It never made any noise? (Second time I asked.)

A. *It never made any noise at all.*

Q. There was no violent action by this animal?

A. *No, not like we were giving it.*

Q. The bullets had no effect?

A. *No, I know we hit it. The three of us were not bad shots, and I know some of the bullets hit. It didn't react, you know. If I had been older, I would have gone over and checked things out more closely, maybe even followed it.*

Q. How old were you at the time?

A. *Well I'm 27 now, so I would have been 13 or 14 years old then.*

Q. All three of you ran away?

A, *Yeah. When we turned and started to leave, it did the same thing in the opposite direction. It was strange, I never saw anything like it before.*

Q. Seen anything similar since?

A. *Nope.*

I can find little reason to doubt Pete's story. Judging from what people who knew him have told me, he is a hard working man who likes to spend his spare time hunting and fishing. There are, however, certain things about this report that bother me. I know a .22 caliber rifle would not in most cases kill a bear, a moose or in this case a Sasquatch. But they would have hurt pretty bad, so I can't imagine why such a creature would just stand there while these bullets were striking it all over. It could be that Pete was wrong, and that in their fright most of the bullets were missing the target. I did ask him later if he could actually see or hear the bullets striking the animal. He told me no, they were too far away from it at the time. Since all three boys were intent on leaving as fast as possible, they did not return to check for footprints, or see if there was any blood around. Was it a Sasquatch they saw that day? Or was it something else like a bear. During the interview, Pete made the statement that bears don't lean against things like trees standing up. He is wrong here. Both black and grizzly bears have been known to stand up on their hind legs to get a good back scratch against a tree. Grizzly bears often have a favorite scratching tree which they will use over and over. However, no bears walk for long distances on their hind legs. The creature Pete and his friends saw apparently could. So what was it these three boys shot at that day? A bear, a foolish man in a costume of some kind, their imaginations, or a Sasquatch?

Having researched the Sasquatch mystery since the late 1970s, I have received a number of reports secondhand. The problem with secondhand reports is the fact that unless names, addresses, and phone numbers were recorded, there is no way of checking the story out. Guy Phillips, a researcher from Winnipeg, Manitoba, wrote to me about a Sasquatch sighting that took place on Vancouver Island near a place called Scott's Falls, in the summer of 1975. Guy also sent me the information concerning the Sasquatch that was seen near Lake Louise in 1984. (See page 114 and 115.) Guy's letter regarding the Scott's Falls sighting follows:

> Found another B.C. item that may be of interest to you. I don't think I mentioned it, as always, you are welcome to it. It took place in 1975, during the summer, on Vancouver Island at Scott's Falls (could be Scutz or Skutz). I went there but like an ass neglected to record it. A guy named Julius Szego and his cousin Ed, about 15 years old at the time, saw a tall creature about 7 feet tall crouched at the edge of the water pulling out

roots. It was on the same side of the falls as they were. It was brownish black in color and had mossy hair on the body. It had no hair on the face and had a flat forehead. It had a human-like face and when the boys had watched it for about 30 seconds, they threw a rock near it to grab it's attention. The creature stood up, looked at them, grunted, and walked off. They were about 40 feet from it and it was near the edge of the bush. This was about 8 p.m. They ran down and alerted the R.C.M.P. and Rangers who came back and took a look around with little result. I spoke to Julius in 1980 and spoke to a couple of his buddies who believed him, when I was out there. People in the area said there was a lot of wildlife around: lynx, puma, which were supposed to be long gone from that part of the Island.

I later contacted John Green, to see if he would be interested in trying to locate just where Scott's Falls was, I could not find it on any of my maps, and John wrote back to me telling me he could not locate the site either. I did realize that a waterfall would have to be a well-known tourist attraction to be mentioned on any road map. I wrote back to Guy, informing him of our difficulty in locating the site. He replied with a photocopy of a road map with the area in question highlighted. Apparently Scott's Falls is located on route 18, north of Victoria between Duncan and Lake Cowichan. A very beautiful part of Vancouver Island, but with a high human population around it. I didn't want to let his report hang there, so I wrote a letter to the R.C.M.P. to see if there was any record of this incident. After about a month, I received a short reply stating that since it was now 13 years later, any file report would have been disposed of by now, and no officer currently posted at this detachment could remember the incident. However, they all had a good laugh when asked.

I also received over the telephone a report from a Mr. James Benson (not his real name), who told me about strange footprints he found while on a fishing trip with his father near Cranbrook B.C., in July, 1975. James had wandered a little way down the bank of the Joseph River. He had to move into the treeline as the river bank - due to high water - lost part of it's shore for about 100 yards. As he walked through the trees, his eyes widened with amazement. At his feet, clearly visible on the wet ground were large, human, barefoot, tracks. He told me that the prints each had five toes and just generally appeared to be human footprints but for one thing. They were about 15 inches in length. I asked him how deep in the ground were the prints imbedded. He told me about an inch. I asked what the weather was like that day, to which he replied, "Overcast and dark." It had rained for three straight days previously and the ground was loose. He followed the prints which traveled down to the river bank where the creature appeared to have stopped and squatted by the running water since there were two impressions side-by-side, toes pointing toward the river. James then yelled to his father to come and have a look at the footprints, but the old man didn't show any interest and told his son it was time to leave, and to get his butt in the car. He did not have a camera, so the tracks were not photographed. Riding back with his dad toward Cranbrook, James pleaded with his father to go back and have a look at the footprints. His father, a man of little patience just said, "What the hell is so important about some animal tracks anyway, it's not like you found bigfoot tracks or something," then he laughed. James decided to keep his mouth shut after that.

I had another report given to me over the phone by a woman named Betty, (not her real name) who told me that she and a friend saw what she thinks was a Sasquatch very close to the town of Hope during the summer of 1974. As a matter of fact, the spot where the alleged sighting took place is inside the town limits. The two girls were with their boyfriends. They had spent the day eating, necking, and throwing a frisbee around. Later in the evening as the sun started to go down, the two girls decided to go down to the bank of the Fraser River. The two boys stayed up by the car which was parked in a place called "Fraser Log PL". The two girls walked down a path to the river's east bank, where they just talked and threw stones into the water. Suddenly, Betty noticed a figure, she at first thought it was a man sitting on a large log watching them. She turned to her friend and asked, "Who's that?" The other girl froze, pointed and cried "My god, that's a Sasquatch!" The two girls then started yelling up to their boyfriends who were by the car. In the meantime, the creature had stood up and started to walk away in the opposite direction. The girls did not stick around to see where the thing went, for as soon as it stood up, they were running back up the trail toward the car.

"What's the matter?" was the first thing Betty's boyfriend asked. Out of breath and very excited, they replied, "We just saw a Sasquatch!" The boys smiled, one of them said, "Yeah, right", while the other started walking about acting like an ape. "They didn't believe us at all", Betty said. The girls, needless to say, didn't speak to the boys again for the rest of the night, "They were real jerks about the whole thing", Betty complained to me. She couldn't tell how tall the creature was, but she did say that when it stood up, it appeared taller than anyone she knew. She mentioned that the thing was hairy and appeared to be brown in color, and that they watched it for about 30 seconds before running back up the path. She also said that although both girls were frightened by the creature, now when she thinks back to the incident, she gets the impression that the thing simply didn't want to have anything to do with them.

Ruby Creek is a small community on the north bank of the Fraser River. It lies along Highway 7 (Lougheed Highway), between the towns of Hope and Agassiz. Ruby Creek is well-known to all people who follow or read about the Sasquatch mystery. It was here that what is known as one of "the Sasquatch classic's" took place in 1941. I am referring to the famous Chapman incident, where Mrs. Jenny Chapman ran from her home with her children, when a creature which fits the description of the Sasquatch approached the family home. I will say no more about this incident as I am making an effort to avoid repeating the classic tales which can be found in previously published books. However, there was another reported Sasquatch sighting at Ruby Creek that is not mentioned in any book. Ralph and Jennifer are a native couple who live in a pleasant home off the Trans Canada Highway near Yale. This story really starts just after I had studied the alleged Sasquatch tracks along the Cow or Cattle Creek. I was back in Hope in a local cafe on the night of August 30th, making a nuisance of myself asking people about the Sasquatch, and getting a lot of strange looks. A fellow named Sonny walked up to me and asked what I was doing looking for the Sasquatch. I told him how fascinated I was with the mystery and tried to explain to him, the same way I've tried to explain to my ex-wife over the years, why I bother to chase after this thing. He became very interested and right out of the blue said "My brother-in-law saw one back in the early 1970s on the highway by Ruby Creek". That was all I needed to hear so I asked him to tell me about it. Sonny went on to tell what he could about the incident, how his brother-in-law still says it was a Sasquatch, after all this time. Sonny gave me his brother-in-law's phone number, and I called the next day. His wife, Jennifer, answered the phone and told me it would be okay if I came over to talk about the incident. When I arrived at the house, I was told that Ralph would be back in an hour or two. This gave me a chance to interview his wife alone and later I could compare the two stories to see if they coincided. It was then that I learned that Jennifer did not actually see the creature, even though she was in the car with her husband at the time.

JENNIFER'S VERSION:

> Ralph, the kids, and I were at a triple feature at the Chilliwack Drive-in. We were coming home along Highway 7, past Kitty Corner toward the Ruby Creek bridge. It was late, between 1 and 1:30 a.m. on a real nice star-filled night. We were approaching Ruby Creek bridge when I leaned over the front seat to cover the kids, who were sleeping in the back of the station wagon. Suddenly I heard my husband cry out "Holy shit!" and at the same time he hit the brakes and brought the car to a sudden halt. I didn't say a word, I just looked out the windshield to see the deer or car Ralph had braked to miss. The kids were now wide awake and asking what was going on. "Did you see that?" was all he said to me. "No, I was covering the kids, see what?" He did not reply, he just drove up the highway a short

distance and did a U-turn. He turned white as a ghost as we drove back very slowly. He asked me again.

"Did you see it?"

"No, I was covering the kids".

"There was something there. It was big".

My husband is a logger, a hunter, and spends a lot of time in the bush, and doing that all his life he is very good at recognizing game. He is not easily frightened.

I then asked him "What did you see?"

"I don't know, but it was big".

"Was it a bear?"

He looked around for a few minutes with the car headlights. I don't know why I did this, but I reached around and locked all the car doors, Ralph still had not told me what he saw. When we started heading toward home again, I asked him again what it was he saw.

"The only thing I can think of is that it was a Sasquatch".

"Maybe it was a bear standing up".

"Not that big, besides I know a bear when I see one".

The next day we went back to look for footprints but we didn't find any. The ground was all covered with loose rock.

After my interview with Jennifer, she made me a cup of coffee and told me that living in this part of B.C. all her life, she was often told stories about the Sasquatch, but she never really believed in their existence. I asked since she never really saw the animal that night if she still had doubts about the Sasquatch. She sat and thought for a moment before replying, "I don't know". When Ralph came home, he was eager to talk about what he had seen that night.

RALPH'S VERSION:

I have been hunting all my life, and I have seen a lot of strange things, but nothing could compare with the Sasquatch I saw that night on the side of the road. I have heard a lot of stories about the Sasquatch, living where I do, but I never thought I'd ever see one. I always thought, if I ever saw one of these things, it would be in the bush on a hunting trip or something like that, not on the side of the highway watching my car go by. Like my wife probably already told you, we were driving home from Chilliwack, we were at the drive-in. It was late, I'd say between 1 a.m. and 1:30, along Highway 7. I had an uneasy feeling like I get when I'm out hunting sometimes. Hunter's feeling, I can't explain it. The hair on the back of your neck stands up, you know. Anyway, we were approaching Ruby Creek bridge, the kids were asleep in the back of the car. Jennifer was covering them with a sleeping bag, and I was driving along, when I saw this huge creature standing on the left side of the road, looking at us. It didn't do anything, it just stood there and watched the car go by. I don't remember what I said, but I slammed on the brakes and brought the car to a halt. Lucky for me, it was late and there were no other cars on the road, or I would probably have been rear ended for sure. Jennifer didn't say anything, she just sat there wondering what the hell I was doing. The kids were now awake, they were pretty upset. I asked her if she saw it. She told me no, she was covering up the kids, so she was looking the wrong way. I then did a U-turn and went back to where the creature was standing, but it was gone. I searched the area with the car lights. I didn't have a flashlight with me. The creature probably moved back into the trees the moment I hit the brakes. I don't think it crossed the highway after I drove by, but then again it might have. It was so big it probably could have crossed the highway in two or three strides. The next day, Jennifer and I went back to the area to look for tracks, but we didn't find any. We looked everywhere, but the area was too rocky for footprints to have been left behind, Nobody believed me when I told them about what I saw.

Later, when I asked Ralph if it might have been a bear he saw that night he replied, "I really don't care what people say, I know I saw a Sasquatch that night".

The similarities between Ralph and Jennifer's version of what happened that night is impressive. Just the opposite from Ann and Jim Smith, who claimed to have seen a Sasquatch at Bears Paw, Alberta (see page 54). They too were interviewed separately and their two stories were compared afterward, and as the reader undoubtedly noticed, I pointed out many discrepancies in the two accounts. In this case, however, there are no discrepancies between Ralph's account and Jennifer's. It is too bad that Jennifer never actually saw the Sasquatch, if that was indeed what it was. Maybe she might have noticed something her husband had missed. He only saw it for a brief moment as it stood there watching the car drive by, but it was a moment that Ralph will never forget as long as he lives.

1974 seemed to be a good year for Sasquatch reports. Not so much by the number of incidents reported to investigators, but by the quality of some of these alleged sightings. I have the interviews on file from the year 1974 which are impressive in their detail and the sheer commitment of the witnesses that they had seen what they claimed to have seen. The first alleged sighting was reported to me by a man who now lives in Calgary by the name of Barry. During the last week of October, 1974, Barry was out hunting with his uncle, a few miles southeast of Mackenzie, B.C. The two men had separated. Barry was hunting along the top of a hilly ridge, while his uncle stayed below to hunt in the valley. Barry had only been up there a short time when he noticed, at a fair distance, an object which he thought was a large black stump of a dead tree leaning against the trunk of a large pine. He didn't pay that much attention to it and kept walking slowly toward it looking for bear signs. When he looked again he suddenly realized there was light showing between the stump and the large pine tree, as though the stump had moved slightly to the right. It puzzled him, and as he stood there he thought, "well maybe since it is a windy day the pine tree had moved with the wind".

He continued walking toward the stump this time not taking his eyes off it. Suddenly it moved again, now Barry was close enough to the thing that he could see this thing was not a tree stump but maybe a bear. He took his rifle from his shoulder and held it at the ready. He took another few steps and froze when it moved again, this time it took a step behind the tree. Barry now feeling a little nervous, realized this thing was not a bear. It was too big to be a man, so what was it? He now decided to warn the thing to back off, so he aimed his rifle above it's head and fired. When the rifle went off, the creature turned and ran straight down the steep hill at a very high speed. Barry who used to be a track runner himself was very impressed by the creatures speed and grace. It ran straight down, not on an angle like a man would do to keep his balance, but straight down at a very steep angle. It ran upright on two legs. Barry ran to the spot by the large pine where the creature had been standing, but by the time he got there the thing was out of sight. Rushing to tell his uncle, he felt a little nervous. He did not know if his uncle would believe him. To his surprise his uncle simply asked to see the spot where it ran down the hill. Later they decided to stop at the Mackenzie detachment of the R.C.M.P. and report what Barry had seen.

When I met Barry, it was obvious that he had led the life of an athlete. He is about 6 feet 3 inches tall and in excellent physical shape. He came to my home to talk about the Sasquatch he saw, and he answered my questions without any hesitation.

Q. Where did this incident occur?

A. East, southeast of Mackenzie, B.C.

Q. What date did this take place to the best of your memory?

A. I had just come out of the arctic so, I think it was the last week of October, 1974.

Q. Was it night or day?

A. It was the middle of the day.

Q. Time?

A. *It would have to be around 2 o'clock, bright sunshine, clear, very windy.*

Q. Describe the area in which the incident took place?

A. *Ummm, we were up on the fire tower road, overlooking the valley. We were up, oh, I don't know, 1,000 feet anyway.*

Q. How far were you from this thing when you saw it?

A. *The first time?*

Q. Yes.

A. *Well, when I first saw it, I thought it was a stump. I was about, oh, 5 maybe 600 feet away.*

Q. What was your first reaction?

A. *Well, when I realized what it was, I was amazed. Especially when it moved.*

Q. What was it doing?

A. *Just standing there, looking at me.*

Q. Was it upright on two legs?

A. *Well, it was standing upright. It looked like it was shifting from one foot to the other.*

Q. Did you ever see it go down on all fours?

A. *Never.*

Q. Was it covered in hair?

A. *Yes. Black hair.*

Q. How tall was it?

A. *Well, when he finally turned and ran, I was about 200 feet away. I'm 6 feet 4 inches, and he was a lot bigger than I was.*

Q. What would you estimate it's weight to have been?

A. *Oh, I can't tell you that, heavy.*

Q. Did you see any facial features?

A. *No, I could see it's eyes though.*

Q. Describe them?

A. *Mostly white.*

Q. Can you describe it's arms?

A. *Big arms. When he turned and ran, his right arm, ummm. When he swung, he turned to the left to run, his right arm came around, and I bet it was bigger than my upper leg. His bicep was bigger than my leg, and I have big legs.*

Q. Could you see if it was male or female?

A. *No, I couldn't.*

Q. How long did you see it for?

A. *Well, from the time I first noticed it and when I first realized that something was there, up to the point where he turned and ran, ummm, 3 minutes.*

Q. Did it ever make any noise?

A. *No.*

Q. I assume it did see you?

A. *Of course. It had to have been there quite a while watching me. The reason I first noticed it is, when I brought the rifle off my shoulder that's when it moved in behind the tree.*

Q. Did you smell anything?

A. *No.*

Q. Did you report this to anybody?

A. *Yes. I reported it to the Mackenzie R.C.M.P. I ran back to my uncle, he was hunting farther down the hill. I told him what I saw. Wait a minute, we did go back to check for footprints but we didn't find any.*

Q. What did the police say about this?

A. *The officer, well he kind of chuckled to himself and said, "Well, we don't believe in them here". Indifference.*

Q. Did you report this to anybody else?

A. *No. Well, over the years at parties, get-togethers, and stuff, when people were talking about strange things, I would say, "Well I saw a Sasquatch in B.C. in '74". Everybody would have a good laugh.*

Q. In your own words, describe what happened?

A. *Well, we had gone up to the fire tower, which was quite high, and I remember it was a windy day. We drove back down to the first plateau. We were bear hunting. I was walking along the ridge and my uncle went down the hill. I was walking north, kind of looking around. I noticed this thing, it looked like a big black stump, leaning against another tree. As I continued to walk, it moved. I could see light between it and the tree. I thought maybe it was just the wind moving the tree back and forth. As I got closer, I thought I saw this thing - stump - shift again. I brought the rifle off my shoulder, thinking maybe it's a bear, and that's when it suddenly moved back in behind the tree. I realized at once that this was no stump, or bear. Rifle at the ready, I again started walking toward it and it moved again, farther behind the tree. I stopped and fired one shot over it's head. When the rifle went off, this thing turned and ran. It was moving very fast, straight down the steep hill.*

Q. Was it on two legs the whole time?

A. *Yes. I was depressed because I was a runner and there was no way that I could run as fast as that thing did. It was taking some pretty long strides. As the thing was running, it hit me that I was looking at a Sasquatch.*

Q. What did you do after it was gone?

A. *I ran up to the spot by the tree where I first saw it, looking down over the other side of the ridge, down the hill. I couldn't see it anywhere, ummm, one thing that surprised me was that there were no slide marks, so that meant that this thing just ran down that hill. There were no places where he skidded or fell, or slid like a man would running as fast as he could down a steep hill.*

Q. What's your uncle's name?

A. *Dean.*

Q. Does he live in that area?

A. *No, he lives in Moose Jaw now.*

The only thing that strikes me as odd in Barry's sighting was the fact he and his Uncle Dean did not find anything at the spot where the creature ran down the hill. I asked him if the hillside was also tree covered. He told me that there was the odd tree, but most of the hillside was dirt, and covered with small rocks. It would seem to me that even if there were no footprints, there would have to be at least marks where the animals feet contacted the ground. Why would Barry make up such a story? He wasn't seeking attention. He asked me to keep his identity to myself, that's why I only mention his first name, as well as his uncle's. It seems pretty clear to me that this was not a bear, deer, or some other known animal. Could it have been someone playing a joke in some sort of costume? I did ask Barry if anybody knew he and his uncle were going bear hunting at this particular location that day. His answer was no. To suggest that it was someone in some sort of costume just hanging around waiting for some gullible person to pass by in the middle of nowhere is just too ridiculous to even consider. We are faced with two possibilities here, either Barry made up this whole story, or he really did see a Sasquatch in late October 1974. If the later is true, I wish he had shot the thing that day and ended this whole Sasquatch mystery once and for all.

About the same time Barry and his uncle were shooting over a Sasquatch's head near Mackenzie, another alleged sighting was taking place several hundred miles to the south, outside the town of Princeton, B.C. Princeton is a quiet little town located just off Highway 3. At the south end of town lies the Similkameen river. The town's built up area is on the Similkameen's south bank. On the north bank there is very little development, at least that was the case the last time I was there in 1987. If you drive up Princeton's main road through the downtown core, you will come to a small bridge which crosses the river, and you find yourself having to turn left or right as the road ends there.

Glen Jackson (not his real name) was jogging with his brother around 9:40 in the morning, late October, 1974. As the two boys were approaching the bridge, they saw what Glen thinks was a Sasquatch, walk up onto the road from the river bank just beyond the bridge. I interviewed Glen when he lived in Calgary, Alberta about the creature he saw as a teenager. Glen is now married with three children and lives a quiet life. He is an avid football fan, with a good many friends in the community. Unfortunately Glen and his family have since moved and I have lost contact with him. When I interviewed him about his sighting, I found him to be easy going and honest. His wife, when asked of her opinion about her husbands story, replied "I never believed in monsters, but if Glen say's he saw it, then there's got to be something to it". I found Glen's story fascinating in it's detail, and at no time did he contradict himself. Here in his own words, is what happened.

> At about 5 o'clock in the morning, my brother and I went jogging in that area. There's a roadway that goes around the airport. We had just gone around the airport - it's about 3 miles

around - and we were coming down toward the bridge. About 200 feet before the bridge we'd had enough of this jogging nonsense, and decided to go home. We saw this thing step out onto the road just beyond the bridge. It's distance from us would be about 150 yards, it wasn't too far. It was crossing the road at the time, and we didn't know what it was. It was black or brownish black in color. We saw him lift himself off the roadway up onto...I don't know, if it is a service road or whatever, but it only took him one step to get up it. He just lifted his whole weight up and on he went walking up the roadway. Now, at that time, I was really scared because it was huge, eh, I was really scared and by that time we had reached the bridge. We were just standing there and as far as we're concerned, it had not seen us yet. It had walked up the road a ways and my brother, I didn't know what he was going to do but he just yelled at it and it turned it's body around to look at us. It watched us for only a few seconds, then it turned around and walked back up the roadway. I don't think it was scared or anything. It didn't act like we were any danger to him at all. Myself, I was really scared. My brother, it didn't seem to bother him, eh. It took my brother about 10 minutes to convince me to follow it up the hill a way. I didn't want to get involved with it because when it turned around, you would swear that you were looking at a gorilla, or something like that. We decided we would try to follow him for a little ways, but he was far ahead of us. We were going from print to print in that skiff of snow. We could see it's prints, so we followed it, eh. The stride, I had to give a little hop to get from print to print. I would say the stride was anywhere from 4 1/2 to 5 feet from step to step and "it" was only walking! When it went up the hillside, up that roadway, there are a few overhanging branches there. It brushed up against these with it's head. I tried to touch them, they must be at least 7 1/2 to 8 feet off the ground. Later we slowly went to the top of the hill, but we never saw it again.

Q. What date did this incident occur?

A. Late October, 1974.

Q. Can you remember the exact date?

A. *No, sorry.*

Q. Was it at night or day?

A. *It was 9:40 in the morning.*

Q. Was it still dark?

A. *No. It was light, lightly snowing at the time.*

Q. Describe the area in which the sighting took place.

A. *The spot is Princeton, just across a wooden bridge. The town is on the south side of the bridge. The thing I saw was on the north side where there is a road along the river that runs east and west. There are some houses spread out along the road. The place is called Allison Flats. You have a cliff there that goes, I don't know, maybe 300 feet, and there's the airport up top. Just a municipal airport, eh, not much to it. It wasn't even paved at the time, I think.*

Q. You saw this thing on the other side of the bridge?

A, *Yeah. On the west side.*

Q. What was your first reaction?

A. *I was with my brother. Ah, he's really gullible so he wanted to go right after it. I was really scared. I didn't know what to expect, eh, because it was pretty big.*

Q. What was it doing?

A. *Well, we had just gone around the airport, eh. There's a path around it. We always jog there in the morning. We had just come down on the east side of the bridge and we were walking the last 200 feet, we were tired, eh. We were just walking and we saw something crossing the road on the west side of the bridge. We stopped. It only took about four steps to cross the road, and one step to get up that little ridge there. He just lifted himself right up on that roadway and started to walk up the path there.*

Q. Did it walk on two legs?

A. Two legs, yeah, even when it went up that one step, it never dropped down on all fours.

Q. Was it covered with hair?

A. Yes.

Q. What color was it?

A. I would say, brownish black.

Q. How tall do you think it was?

A. We estimated about 7 1/2 to 8 feet because when it went up the hill, it was just brushing branches that were overlapping the road. They were hanging off the bank side there. I couldn't touch them, so I think the thing was 7 1/2 to 8 feet tall.

Q. How heavy?

A. He was heavy, for sure. I don't know 400, 500 pounds, who knows.

Q. Did you see any facial features?

A. When he stopped, my brother yelled out to it. He said, "Hey!" It turned around and it was like you were seeing, I don't know, King Kong or something. It had an ape-like face.

Q. Can you describe the eyes or nose?

A. The nose was not protruding much, the eyes were really dark, eh, but I couldn't really tell.

Q. Ears or mouth. Did you see it's teeth at all?

A. No. I wouldn't say it had a real hairy chest. We were debating through the years, whether it was male or female because it did have all fur around the chest area. Unless it was muscle. I don't have any idea, but it was different, it wasn't just fur.

Q. Could you describe it's arms?

A. The arms hung quite low, past the knee area, very big and covered with fur.

Q. How long did you see this creature for?

A. Three minutes, if that. It wasn't worried about us at all. It didn't run. It didn't see us until it crossed the road and was already on the bank heading up the roadway.

Q. Did it make any noise?

A. No.

Q. It did see you?

A. As far as I'm concerned, it didn't see us until my brother yelled at it.

Q. How did it react when your brother did yell?

A. It might have been slightly startled, but he turned and

we could see it wasn't a gorilla costume or anything like that.

Q. Did it smell at all?

A. No.

Q. After it was gone did you check for footprints?

A. We did. It took my brother about 10 minutes to convince me to follow him. He wanted to follow him right off the bat, but like I said, I didn't really want to get involved with it. But when he headed up around the corner, about 10 minutes later we headed up to the same place, and there were footprints. My brother tried to preserve one by putting cardboard over it. It was snowing at the time, eh. There was just a skiff of snow on the ground and that's what the prints were in. We could see that the print was bare, about 13 inches long.

Q. 13 inches?

A. Yeah.

Q. Was the print very deep?

A. No. The road there is fairly hard, and winter was coming. The prints were in this fresh snow.

Q. What's your brothers name?

A. Alan.

Q. Did you report this to the police?

A. No.

Q. Did you tell anyone?

A. Just family members.

Another very detailed story in which the witnesses want to remain anonymous. If this sighting is true, I can't personally see how it could be a case of mistaken identity. Could they have been the victims of a hoax? Glen doesn't think so. He described how the creature's head brushed branches that were 7 1/2 feet off the ground. Also, the animal lifted itself up the side of the road embankment in a way that no human could, according to Glen. So what is the answer? I think I will play politician here and sit on the the fence. If the Sasquatch does indeed exist, Glen and his brother Alan saw one in late October, 1974. If the Sasquatch does not exist, the whole story is nothing more than a clever piece of fiction. It is up to the reader to decide what he or she thinks.

Most Canadians anywhere in this vast nation of ours, live in large, or semilarge urban areas. If you were to ask the average person off the street in Victoria whether he or she had ever seen a moose, elk, or deer, they would likely say, yes.

If you asked them where they saw it, the odds are pretty good that in most cases, the animal was seen either on the road or beside it, while the person was driving by in their car. Sightings of the elusive Sasquatch seem to fall in the same pattern as any other animal. If you look at a detailed map showing the location of reported sightings, you would notice that the vast majority seem to be close to roadways, rivers, lakeshores, and campgrounds. The Sasquatch, if indeed it exists, is no doubt a creature of the deep forest, avoiding contact with humans just as any other animal would. If the Sasquatch is nomadic, it would in it's travels while searching for food, occasionally come close to human beings. Avoiding people if it can, it is no doubt as curious about us as we are of it. B.C. has had it's fair share of roadside incidents, one of which I have already covered in this chapter. In this case, the sighting was very brief while the driver was left to replay the incident over and over in his mind. What would you (the reader) do if you were driving on a public road and something large, hairy, and walking upright came out of the treeline in front of your car?

Would you report it? Or would you keep the encounter to yourself. If you chose the latter, you would be with the majority. Sightings of this nature occur more often then anyone realizes. Only a handful will be reported.

Randee Ford is a young man who told me of a strange animal he saw while he was driving a newspaper truck from Cache Creek to Prince George. He would make the trip four times a week, driving all night, dropping newspapers off in towns as he went along. One summer night in 1978, about 3 o'clock in the morning, Randee and a female coworker named Millward were driving along Highway 97 just outside Quesnel, B.C. when he noticed something off to the right-hand side of the road in the ditch. The highway had a sharp bend to the left at this point so the area was illuminated by the truck's high beams. The creature startled by the lights stood up to look at the oncoming truck for a second, then turned and headed up the roadside embankment, running. The creature was covered with brownish black hair and it ran upright on two legs.

Occasionally, it would put it's hand on the ground as it ran up hill. The creature appeared to be 8 or 9 feet tall and it weighed about 700 to 800 pounds. It had an ape-like face according to Randee. The sighting was about 45 seconds in duration. He did not report this to anyone. He told a few friends about it over the years, and in most cases was met with skepticism. I asked him if he could be mistaken and had seen a bear, but he told me that he had asked himself the same question over and over again and, in his opinion, it was not a bear. The woman with him was badly shaken by the experience, so he did not stop to try and see the animal again, or to look for footprints.

In late June, 1982, I was contacted by a man named Robert Harrison who told me about an unusual animal he saw while he and another man (Frank) were driving along in a '78 Chev 4-wheel drive truck. It was about 6 o'clock in the evening, the sun was still up, it had been a clear sunny day. The two men were talking when they suddenly noticed a creature standing in the middle of the road. It was covered with reddish brown hair, and it stood upright. The creature appeared to be watching something to the left. When the truck came into it's view it turned it's head to look at it, then the animal turned right and started to run down the road ahead of the oncoming truck. It only did this for a short distance, and the two men were amazed when the thing leapt off to the left side of the road, and disappeared in the pines. Robert slammed on the brakes, leapt out of the truck and tried to get another look at the thing, but it was gone. The two men did look for footprints but did not find any, just some scuff marks off to the left down the embankment where the creature leapt. Robert guessed that the creature was just over 6 feet tall. He also said the thing did not seem to have hair on it's face or hands. The two men were going to report what they saw to the R.C.M.P. but decided against it.

I got another phone call from a Mrs. Agnes Perkins of Calgary, age 65, who told me that she was driving through the Rogers Pass on the Trans Canada Highway with a friend (Charlotte White) on the 18th of August, 1987, when they saw something strange about 800 yards ahead.

What she described was an ape-like creature covered with black hair, standing on the highway's right lane. She thought it stood about 7 feet tall. As the car got closer, the creature turned around and started climbing up the steep hillside to the right. It only put it's hand down a couple of times for balance. According to Mrs. Perkins, the rest of the time it was running on two legs. The two women did not stop, they just drove on.

On Monday, May 2nd, 1982, I received a phone call from a man who saw a newspaper article about my research into the Sasquatch. The gent would not give his name and told me he did not want to get involved in any investigation, or have anything to do with the Sasquatch. He then went on to tell me that it took some thinking before he finally decided to give me a phone call and tell me what he saw. I then told him I didn't usually take reports over the phone, especially if the witness won't identify himself or give me a phone number or address. He then reluctantly told me his name and phone number, telling me to hang up and phone him right back. This I did. He then made me promise not to publish his name or repeat it to anyone, so I agreed. This anonymous fellow told me he was driving with a friend (he would not give me his friend's name), north on Highway 95A, toward Wycliff, B.C. on the night of April 28th, 1988. They were approaching the bridge across the St. Mary River when they saw a large, black, hair covered creature come out of the trees onto the road in front of their car. The creature did not stop, it just walked at a fast pace across the highway and disappeared into the forest on the right side. The two men did not stop but drove on to Wycliff trying to decide if they should report this. They decided not to. It was about four days later, in Calgary where he lives when he saw the article in the Calgary Press and debated for awhile about whether or not to phone and tell me. He had no idea how tall the creature was that he saw, but he did say it was taller then anyone he knew. When I asked how heavy it was, he told me "I sure as hell wouldn't want to wrestle it!"

Still from the famous Patterson/Gimlin film shot on October 30, 1967, at Bluff Creek, northern California.
Photo: © Rene Dahinden/1968

Now, despite all the testimony from witnesses who insist that it could not have been a bear or someone in a costume they saw, hoaxes cannot be ruled out. I recall a Sasquatch sighting along Highway 7 near Mission B.C. by a busload of witnesses. It was a Sunday in May, 1977, and this sighting received nationwide publicity. Even the police were involved. Rene Dahinden still laughs when he thinks back to the police taping off the area where footprints where found, covering them with a tarp. Two big mounties guarded them while dozens of sightseers watched. "Oh, ah, look at that", camera shutters clicked as the tarp was rolled back to reveal the tracks.

Later, Rene Dahinden and John Green said publicly that there were certain things wrong with the story - they had suspected a hoax.

Three pranksters soon revealed themselves and their elaborately planned prank. I have studied photos of the costume they used, and I still can't see how anyone with any sense could have been fooled by it. The human brain has been known to play tricks on one. People can think they see things when, in fact, they are seeing something completely different, especially when you only get a glimpse of it for a few seconds. I had a personal experience with visual misinterpretation, as I like to call it. In the summer of 1988, I was driving along a forestry road from Lillooet, west to Pemberton with my ex-wife, Pam. This was the first and last time she accompanied me on a Sasquatch research trip. We were driving along watching the sides of the roads for anything that might appear like footprints.

Suddenly she said "stop". I asked her what she saw. "I saw something big and black walking up there". On the right side of the road there was a river, beyond that a steep hillside with several large burned out black tree stumps. There had been a forest fire through this area some years ago, and the second growth was well under way, only several large black stumps remained where the fire had been. I was not about to jump out of my truck, cross that river, and start climbing that hill unless I was sure she had seen something. She told me that out of the corner of her eye she saw what looked like a large black object walk between two large tree stumps. I put the truck in reverse and backed up about 500 yards and stopped. Slowly driving forward again, I told her to look in the same direction she did before and tell me if she saw it again.

Sure enough, there it was, the legendary Sasquatch, walking between these large tree stumps. Except instead of a large unknown primate, it was a third tree stump about 200 feet further away with a large knob on top that in the fading light (it was about 8:30 at night) appeared like a head on top of a pair of massive shoulders. With the angle as it was, the fading light, plus the fact that we were moving, the third stump did seem like a large creature walking from behind one stump to the other.

A couple of days later, we drove past the same spot again, only this time it was midday. Bright sunshine and the moving object appeared to be what it really was, a burned-out tree stump. How many of the roadside sightings I wonder could simply be visual misinterpretations? What about the many where the sighting involved a lot more detail? I don't have the answers, but I always keep my eyes open when I'm driving through British Columbia, the traditional home of Canada's hairy giants.

CHAPTER EIGHT
"What Is The Sasquatch?"

The text of this book has, thus far, covered many sightings and possible encounters with the Sasquatch. What is it exactly, though? With the combined effort of other researchers, years of eye witness testimony and the few theoretical possibilities from the scientific community, a composite-type verbal description can he formed dealing with physical characteristics and probable instinctive/habitual traits or practises.

Based on current evidence, it would be reasonable to infer that the Sasquatch is a higher primate, hominoid, which is a classification reserved to describe mammals in the man-like family such as apes, orangutans and humans. Presently there are only five identified higher primate species, the Sasquatch, if proven to exist could possibly be the sixth.

According to eye-witness testimony the animal moves only on two legs and, although it walks very much like a man its stride is nearly twice as long and its speed is considerably faster. Scientists examining a film made of the creature claim that it would be virtually impossible for a human person to imitate the stride and speed, even with extensive training.

The Sasquatch has been reported to be completely covered with black, brown, red-brown, grey, white or silver tipped hair. Black, brown and red-brown coloring is most common with this animal and lighter coloring is usually reported in extreme northern areas of the globe such as Siberia and the Canadian Yukon and Northwest Territories.

The height of the animal has been described as being between eight and fifteen feet tall, the tallest apparently residing in the Alberta area. It is very bulky, weighing anywhere above 200 pounds and very muscular in the arms and legs. Occasionally one is reported to be very thin with a small stature which could be a result of dietary deficiency, age or deformity.

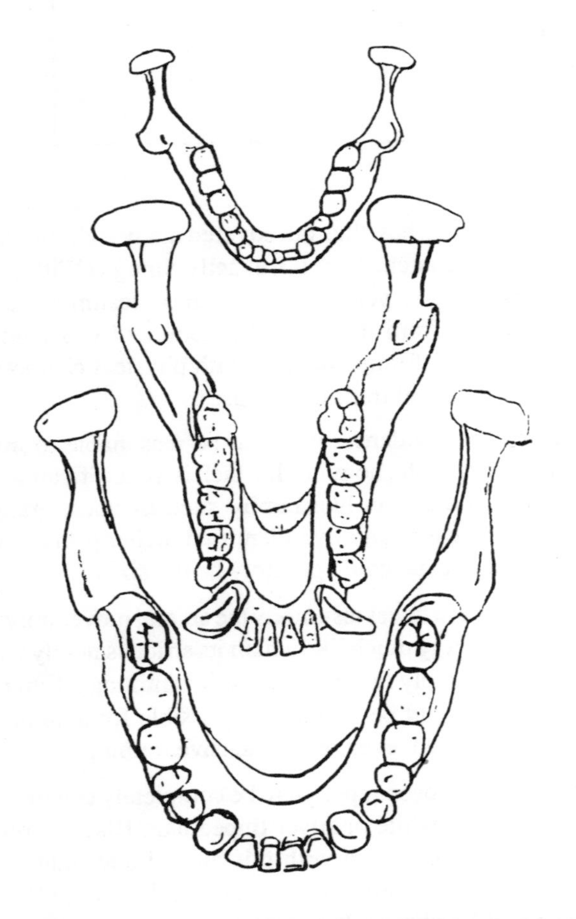

If you compare the lower jaw of the Gigantopithecus (bottom) with the lower jaw of an adult man (top) and the lower jaw of the gorilla (centre), you can see that the Gigantopithecus was huge. Also the spread of the jaw is more like that of the human than the gorilla, whose jaw is much narrower. This argues, in theory at least, that the Gigantopithecus was an erect bipedal hominid (walked upright).

A Sasquatch face is somewhat like that of a gorilla or a very old weathered man. Its mouth is large with big square teeth and occasionally the eye-teeth, when seen are said to be slightly larger that the rest.

The jaw line could possibly be matched to the Gigantopithecus "Kwangsi Giant" (see diagram on page 94), which is extremely close in shape to a Caucasian adult male though more than twice its size.

According to compiled reports (which have been viewed as legitimate), the Sasquatch does not appear to be dangerous or harmful to the human race. In Alberta, the animal has appeared to be either unaffected or fearful of the sight of humans. I know of no account since the turn of the century where the animal was reported to intentionally harm a person.

Older stories from the 19th Century as well as some of the tribal Indian legends convey that the creature is to be avoided. There have been reports of Sasquatch's bluff-charging much like the mountain gorilla is known to do. In these and related incidents the animal veered off before it reached the witness. Assumably, the persons involved in a bluff-charge aged a few decades in the few seconds the incident lasted.

Aside from the above stories, however, the animal remains a possibly shy and even reclusive creature. Some researchers have catalogued it as nocturnal (active only at night).

Habits of the creature, according to compiled reported incidents draw a very crisp picture of its natural instinctive personality. The Sasquatch has been known to scavenge food left on picnic tables overnight yet seem to elude areas lit up by campfires. Some stories recall the animal peering inside tents with the flaps left open. The creature has also been known to shake campers, approach houses and private property, peer into windows and run beside cars (to get a better look at the funny hairless creature behind the wheel). Quite obviously, the Sasquatch has a curiosity that may someday get him into a load of trouble!

The animal seems to have no fear of water and has been reported wading and splashing in rivers and lakes. Taking into account that they can walk faster than a man can run (Olympic runners not included), it is evident that the energetic animal will not be caught until it is good and ready.

Let me, for a moment, take you on an imaginary tour with a Sasquatch, through the foothills of Alberta on an early spring evening. For amusement, we shall call the animal "Stretch".

It is 8:30 p.m., Sunday evening and Stretch has just awakened from a full night's sleep. He moves away from the nest of branches and the pack of family that still lays slumbering on the cave floor, and heads towards the dim light of sunset, creeping through the rocks by the west side of the den. Stretch stinks. Stretch is hungry.

After leaving the cave, he moves through the dense brush with ease stepping from time to time on small bits of broken rock. He approaches the mountain spring and stands beneath the fall, letting the water whip down across his fur and cool his tired body. Next to the spring are four large berry bushes with ripe, red, juicy berries. Stretch leans over to gather a few bushes full (told you he was hungry) then makes his way up an embankment towards a newly hewn cut line.

On either side of the path are animal tracks that Stretch doesn't recognize. They do not begin or end but run up both sides, always parallel. Out of Stretch's instincts dictate him to go farther and discover this new animal in the area. He heads up the cut line, unknown to him, towards the campsite of city-weary outdoorsmen.

The closer Stretch gets to the site of the people the more his nose bothers him. The smell is not familiar and he is increasingly uncomfortable with the sounds coming from just up ahead.

Finally, Stretch sees it! Not one animal but a whole herd! Peering through the bushes, into the clearing, he notices that their paws aren't anything like the tracks they made. Perhaps Stretch is wondering if these animals walk on their heads? Suddenly, one of the animals points its paw in his direction and makes a shrill high scream (mating call maybe?). Stretch lingers for a few moments observing the mass confusion in the clearing then, with an instinct of fear, runs back into the bush, avoiding the track line and steering towards the cave.

Stretch is an animal, so he didn't read the papers the next day nor was he aware of the dissatisfaction of the people with the fact that he didn't stick around to get his picture taken.

Although this is hardly a realistic tale, some of the characteristics of Stretch are similar to believed characteristics of the actual Sasquatch.

The animal is assumed, like other animals of the wild, to have a keen sense of sight, smell and hearing. It is also believed that the animal is vegetarian, although it has been reported as seen eating small rodents. Finally, it is believed that the animal is either extremely curious of humans or genuinely afraid of them.

The idea that the animal is nocturnal stems from the quantity of reported sightings which have taken place in the evening and at night. Those who have reported Sasquatches as being in front of their cars at night state that the animal's eyes shine like that of a cat's shine looking into powerful light. This lends authenticity to the belief that the the Sasquatch has superb night vision.

These, however, are yet theories gathered from eyewitness accounts of an animal who, it is widely believed, may not exist. If the animal is proven to exist, there must be further evidence either uncovered or unintentionally misrelated.

The fossilized remains of a lower jaw belonging to a huge Asian ape, known as the Gigantopithecus and believed to have resided in the area of China and India, may place the Sasquatch into the fossil record. The most interesting thing about this lower jaw is its width, which, when compared to that of a gorilla is considerably wider. The gorilla, whose lower jaw is very narrow, spends most of its time on all fours and its head does not sit on top of its body. The Gigantopithecus, however, has a wide lower jaw very much like that of a human. This has lead researchers to conclude that the Gigantopithecus is an erect bipedal animal (one which walks upright) with a neck. Also judging by the size of the jaw, the animal was incredibly huge.

It is important to note that reports of the Sasquatch describe it as being an upright animal with shoulders and a small neck. Could this giant Asian ape be an ancestor to the Sasquatch? One theory is that the Gigantopithecus moved over the land bridge into Alaska from Siberia at the time the Indians began to inhabit North America. The animals left in China somehow became extinct and those who moved over to North America became known as the Sasquatch. This theory would lend credence to the sighting of similar creatures across Russia and, in fact, various other regions of the world. Unfortunately, aside from the reported presence of more human-like, hair covered creatures (Alma's) north of China, we have no direct evidence that the Gigantopithecus ever travelled.

Yet another piece of evidence which could possibly fit into the study of the Sasquatch is a discovery by Dr. Jerold Lowenstein of the University of California in San Francisco.

Dr. Lowenstein has developed a process using immune reactions to identify proteins, enabling him to determine the possible source of feces, i.e: bear, primate etc. In 1987 Bob Titmus of Harrison Hot Springs, B.C. submitted a hair sample collected from trees. Dr. Lowenstein was able to identify the hair as coming from a primate, yet he could not identify whether it had come from a human, chimpanzee or gorilla.

"He therefore assumed it was human, but in fact it was nothing like human hair, since it had wool hairs as well as guard hairs and the guard hairs were pointed, not cut off at the outer end. Unfortunately he had ground up the whole sample, otherwise it would have been simple to check his three possibilities with a microscopic comparison. I am quite sure all three would be eliminated, as I doubt that chimps or gorillas would have wool hairs either - this is a point not covered in any reference ... except that lowland gorillas definitely have no wool hairs." (John Green, Letter September 1988)

Most definitely a higher primate was leaving hairs on the trees in California, a primate who could not be conclusively identified by advanced methods of identification.

CHAPTER NINE
"A Hunter's BIG Bear?"

On the night of October 20, 1987 an unidentified caller reported that he and a friend had seen what they believed might be a Sasquatch in April of the same year.

According to the man on the other end of the line, he and a friend were on a hike along a cutline a few miles northwest of Rocky Mountain House, as the sun was setting. Both men were apparently hunters and well versed in the identification of animals. Then, out from the tree line stepped a 7 or 8 foot tall, dark hair-covered creature standing on its hind legs.

The two stopped dead in their tracks and stared straight ahead getting a full view of the strange and unidentified, to them, animal. The animal also saw them and bolted back into the tree line and out of sight. According to the two witnesses, the animal ran back on its hinds.

These two men were not the only hunters to have ever sighted a similar creature. Hunters are a unique brand of people. A good hunter is disciplined in animal identification and tracking. Because Alberta's hunting laws are stringent, men tracking down game must be careful to identify the target as an animal he is licensed to shoot. Not just the animal must be identified, but also the specific breed - if the wrong breed is shot in the wrong season a hunter can face hefty fines. However, I have on file a good number of reported sightings from hunters - not the type of report to dismiss as a case of mistaken identity.

The second Alberta Sasquatch sighting in my files took place in late summer of 1954 north of Blairmore and Coleman by Mr. Lidio Orlando, a hunter who lived just north of Edmonton.

Mr. Orlando wrote me a letter to tell of his sighting. Following is an excerpt of that letter.

> "...It was open hill country with mixed spruce, pine and some poplar. The highwood range was to the east and the Rockies to the west. I was out hunting rabbits with a .22 calibre Winchester and my dog King. About 4:00 in the afternoon I was sitting on high rock ridge looking east, when I spotted a tall black creature walking swiftly towards me. I just sat there watching, thinking it was a man at first, but the closer it came the more unman-like it became. Walking erect, it covered ground at an unbelievable rate. No human could possibly run that fast, let alone walk. When I first saw the creature, it was approximately 400 yards away. It walked directly towards me to a point about 180 yards east of me, then turned right and headed northwest up a ravine until I lost sight of it.
>
> ... It walked directly under a large tree, with a large overhanging branch about 16 feet off the ground (and) ... cleared it by about 2 or 3 feet. ...the most striking thing I can remember is the size, and the gait which was unbelievably fast."

What was it that Lidio Orlando saw in the summer of 1954? Could it have been a bear? Bears do not walk on their hind legs. Could it have been someone playing a joke on Lidio? If it were a joke, the man playing it must have been very tall and extremely energetic, beyond any man known in history!

At 4:00 p.m., June, 1974 in Bragg Creek Provincial Park Carl Melnyk a hunter saw what he believes might have been a Sasquatch, while spending the afternoon with friends at the campground.

Carl was sitting on the bank of the Elbow River. Approximately 100 metres away, just inside the tree line, something was moving. The movement caught his attention and he glanced over. A large dark brown or black body covered in animal hair came into view. Naturally Carl thought the body mass to belong to a moose or elk.

Then, Carl's eyes grew wide and he strained to get a better look at what was in the trees; why - because it was walking upright, on two legs, just like a person! He watched in amazement for close to 15 seconds before the animal slipped out of sight. Carl quickly went to the area where he had seen the creature and searched for tracks but none were found. Later he told his friends about the incident - they laughed.

Carl's curiosity was not quenched, however, until 13 years later. When he returned to his apartment in Calgary Carl saw my ad and called. He wanted more information on the Sasquatch - precisely what I had asked for in the ad.

According to Carl the animal he saw was at least 600 pounds, big, bulky and husky with long thin arms. What convinced him that it was not just a large man?...

"After a few seconds I noticed that it was moving on two legs and that is when I realized that it was not a normal animal, at least not a common one. So I just stared at it for about 15 seconds and then it took off in a big hurry ... ripping through the trees like there was nothing there ... it was the speed that convinced me this was not a man!"

On the 10th of July, 1988 three teenaged boys and an experienced hunter fled a campsite, panic-stricken. Later that evening, the R.C.M.P. and the local Warden's office were contacted and given what could have been a "bear-scare" story.

A close look at the report, though, led me to believe that an animal other than a bear caused the panic!

A Lethbridge man had taken three young boys camping and backpacking. After taking possession of a cabin by Upper Twin Lake in Waterton Lakes National Park the foursome headed out for an afternoon of fishing and backpacking. Still within hearing distance of the cabin, they settled by the lake with their fishing rods, bated the hooks and dropped the lines - all hoping for a prize catch. Suddenly a commotion coming from the direction of their cabin broke the afternoon lull. A high pitched scream echoed through the trees - a scream that none present had ever heard.

As mentioned earlier, the leader of the troop was a hunter who had heard wild animal calls many times before but this was not an animal call he could place.

When the boys returned to the cabin their peaceful camp had been disturbed. Something, someone was prowling through equipment and personal belongings. Fearfully, they packed up immediately, and left the area, not checking for tracks or signs of who or what was driving them away.

Sometime later I was contacted by these people. At that time I had in my possession a cassette taping from Fouke, Arkansas, of what is believed to be a Sasquatch call. During our interview the tape was played. All four agreed that the sound was exactly the same as the high pitched scream coming from the direction of the cabin.

Most possibly the scream could have come from a human intruder using scare tactics in order to rob the site or even drive the boys from the cabin. It could have come from a wild animal who was hurt or dying. Why not? Nothing was stolen from the sight - just disturbed; as for the animal theory it doesn't work for the same reason - a wounded animal would hardly just disturb a campsite then quickly disappear from the area.

It would not be presumptuous to assume that if the Sasquatch does exist it has a distinct call or cry. Many reported incidents of strange "screams in the night" unidentifiable as human, bear or any known animal have come from areas frequented by actual Sasquatch sightings. In 1978 a family reported a very similar incident.

The Simms family (alias) were off to spend a day at Jackfish Lake, north of Highway 11 between Nordegg and Rocky Mountain House. Mr. Simms brought the family car to a stop. The family piled out of the vehicle and began unloading the car, setting up day-camp. The area was dense with bush and trees. Birds were peacefully chirping, small creatures were scurrying across the landscape. then, in a split second the calm was broken.

The birds ceased singing and took flight from the area, small critters scampered to safety and one incredibly long, eerie scream rang out through the trees, across the lake filling the sudden silence. Mr. Simms blood ran cold - he had never heard anything like it. The whole family stood staring toward the bush. Something big moved quickly deep into the tree line.

The Simms' threw their belongings back inside, scrambled into the vehicle and drove off. Not one of them would brave investigating.

Mr. Simms has never forgotten the sound of the scream.

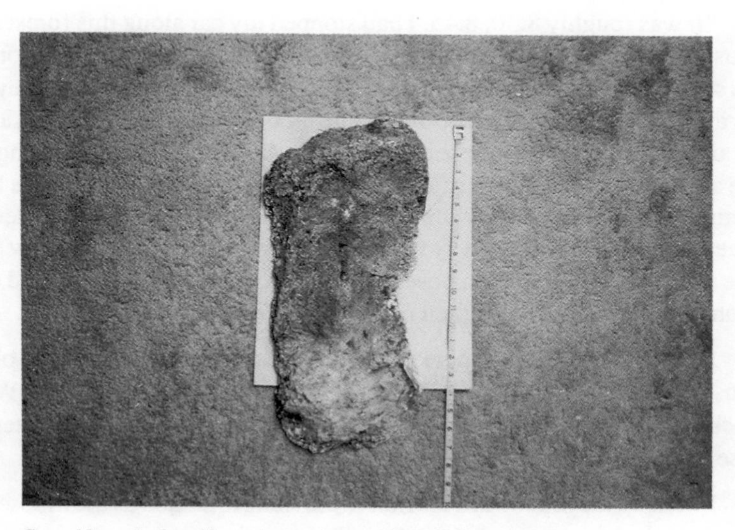

Cast of footprint found by author on August 1986. Chilliwack River.

Mr. Donald B. Kenny was a single shot away from solving the mystery of the Sasquatch once for all.

Donald Kenny didn't answer an ad in the paper, he made no report to the R.C.M.P. or Warden's office but what he encountered, what he had seen with his own eyes during a 1975 hunting trip warranted telling someone —

someone who would take his story seriously. Donald Kenny wrote to John Green, a well known author and authority on the Sasquatch search.

John Green forwarded a copy of that letter to me. I contacted Mr. Kenny and requested an interview.

On November 7, 1975 Donald was moose hunting fourteen miles west of Sundre, Alberta. He fired a shot at the Sasquatch and missed. I asked him to explain exactly what happened.:

"It was roughly 8:30, 8:45. I had stopped my car along this forest road because I'd seen a bull moose in the swamp. It was almost right over on the north side of this swamp. I got out of my car, loaded my rifle and took several steps away from my car into the ditch. The moose was standing broadside to me, I aimed and shot. The moose did fall down. As it fell, its four legs dangling in the air, I tried to reload the rifle again but it was jammed. While I was busy fidgeting I looked up and saw this moose was getting back up on its feet. It proceeded to move rapidly away from me, heading up-hill onto the cow trail. At the same moment that the moose reached the trail a Sasquatch walked onto the same trail. Both animals then disappeared from sight.

I went to the place where the moose had fallen to check for blood because, believe it or not, my primary concern was to get the moose. When I checked and found only a little bit of blood I figured I'd better go after the moose and see if I could get it.

I went up the hill toward the trail in the same direction as the moose and the Sasquatch. After going about 200 yards the tracks became scarce. The area was above the snow and the ground was covered with a spongy moss. I couldn't track the moose very well.

I circled the area in a wide circle, then an even wider circle. I had just finished the second circle and was about to start a third when I had come into a small clearing. The sun was coming in from the east side, I looked east and could not see any tracks, I looked south and still couldn't see tracks.

Then, for some unknown reason (hunter's intuition) I just knew that I was being watched. Slowly turning my head to the left my eyes met the eyes of the Sasquatch. It was only 90 feet away (approximately). Then I felt fear. Taking the safety off my rifle I started to bring it around and the Sasquatch took off.

I thought it was going in a northeast direction, so I aimed for a spot between two pine trees and fired. After this I made another circle to see if I could see it again because I did not know if it had been hit. The circle brought no results - I did not see the animal again."

Mr. Kenny also gave a description of the animal. He claimed the animal's facial features were comparable to that of a human's and covered in reddish brown hair. It walked exclusively on two legs, not once going down on all fours and its arms hung below the waist.

The animal he alleged to have seen that day made no sound other than sounds caused by natural movement. It did not bare its teeth or show signs of aggression.

Mr. Kenny saw what he believes was a Sasquatch. He saw it in broad daylight twice. He saw it clearly and he fired his rifle with the purpose of hitting it. Had Donald Kenny actually gunned down the animal and brought its corpse (or injured body) into civilization the mystery would be over.

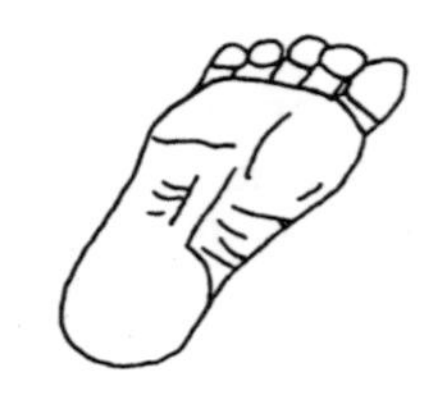

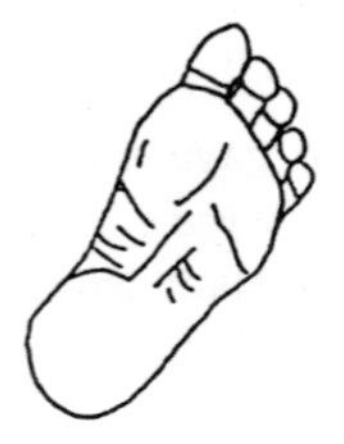

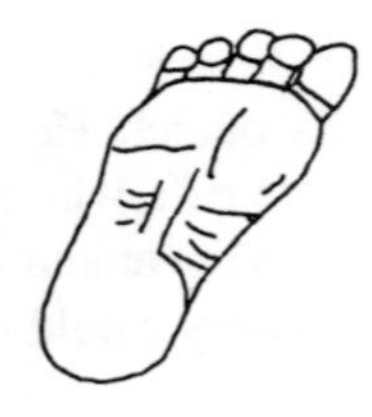

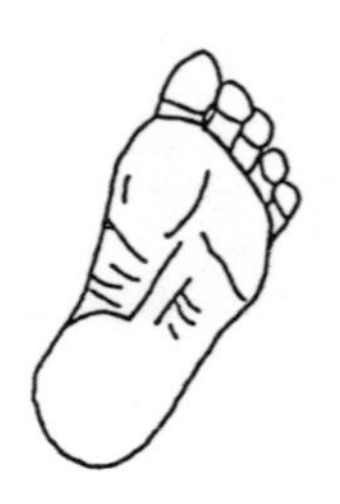

CHAPTER TEN
"The Clearwater River"

From the town of Caroline to the Rocky Mountains lies the Clearwater river area and for the past 15 years it has been a buzz of Sasquatch activity. Unlike most camping and wildlife areas in Alberta, the Clearwater River area is widely used for recreation, camping and wildlife exploring.

Amidst the wild untouched landscape, mazes of dirt roads, cut lines, and occupied campgrounds twist around the area. A drive along the banks would afford the pleasure of seeing wildlife - elk, deer, black bear and sometimes grizzly intermingled with the human trappings of partial civilization. Phyllis Lake Campground, Tay River, Seven Mile Recreation, Elk Creek, Cutoff Creek, Equestrian Campground, Peppers Lake and Ran Falls are widely used by people - yet the animals remain in the area. This could be the reason for such frequent sightings of the Sasquatch; not that there are more Sasquatches in Clearwater River, but that there are more people to see them.

Just outside of the Phyllis Lake Campground on the banks of the River on a pleasant September afternoon in 1976 two men were hunting grouse. Gary Schmidt of Alberta and Peter Garyall of Alabama, U.S.A. walked parallel, Gary on the south bank and Peter on the north bank. Their hunting efforts that afternoon were futile.

Peter walked along until he came to a bend which gave him a clear view of the river. Something piqued his curiosity. A large form about 200 yards down the river squatting by the water, splashing it over its body. "Hey, have you got apes up in the mountains?" he yelled over to Gary.

Gary, caught off guard by the question raced over to the north bank by the bend to get a look at what Peter was calling an ape. He looked down 200 yards and stood watching, stunned. There along the waters edge, not a block distance away was a creature neither man had seen before - it almost looked like an ape.

The creature was about 9 feet tall, covered in black hair, its head was sitting on top of its shoulders, not jutting out from just above the neckline. Its arms were incredibly long and its face had the appearance of a cross between a human and a dog.

The heavily wooded landscape just north and south of Highway 11 is where a high number of Alberta's Sasquatch sightings seem to occur.

For nearly five minutes the animal washed itself and the two men looked down upon it. The shallow, rushing river drowned out any sounds and neither Gary or Peter could hear anything coming from the animal.

Then it looked up towards the bend, spotted its observers and hurried off into the tree line and out of sight. Neither Gary or Peter reported the incident to authorities.

Sometime later, Vladimir Markotic, professor of Archaeology at the University of Calgary who has been researching the Sasquatch mystery since the early 60's landed a spot as a guest speaker on a radio talk show. When the talk line was opened, Gary Schmidt called Professor Markotic and reported the incident.

The public's curiosity cat sprang into action. A few months later I was invited as a guest on Calgary's radio talk show "The Home Stretch" to discuss the Sasquatch. When the talk line was opened Gary Schmidt was on the other line wanting to know more about these creatures. The name was familiar, Vladimir had passed on this man's name and phone number to me and I had called Gary to hear his story.

Gary wanted to know more about the Sasquatch, I wanted to know more about Gary's sighting. I requested an interview. He was not too enthusiastic about being interviewed, however. Gary had been on the receiving end of jokes and jeering because of the sighting. He was, of course, not alone. I explained that many people who previously came forward with a story of a sighting required more courage to face the suspecting and critical public than to face an angry Sasquatch head on. Gary decided then to use his courage and agreed to meet with me.

Q. How much of the creature did you see and for how long?

A. *I saw the whole thing for about five minutes. I was no more than about one city block from it.*

Q. Could you describe what took place?

A. *"I was hunting rough grouse with a friend of mine along the Clearwater River near Caroline. We were walking along when he yelled up to me saying 'you got apes up here in the mountains'? I went running down and I saw this big animal in the river throwing water on top itself.*

It was splashing itself with water. I ran down toward it a little ways and my friend says 'are you crazy? Such a big animal!'. After about five minutes it spotted us and just got up and took off."

After the interview we sat and talked for awhile. Gary told me how he had been made fun of by his friends and some of his family. Later we shook hands and as I walked out the door his last words were "Mr. Steenburg, you know now I wish I never saw that thing... Nobody believes me and so I just don't talk about it anymore."

Another Clearwater River incident took place in the first week of August 1984. The man involved does not wish to be identified publicly so I call him Bud.

Author searching for footprints, Clearwater River.

Bud was camping with his wife at the Seven Mile Recreation Area. As dusk was falling He headed out for a walk to stretch his legs. Bud walked down a cut line lined with trees between 9 and 20 feet tall for about 4 or 5 city blocks.

Tay River Campground, 1979. Vladmir who is almost 6' stands at spot Sasquatch reportedly crossed the road.

Bud developed a strange sensation that he was being watched. He looked around and just inside the tree line He saw the head and shoulders of a creature that must have been about 12 feet tall. It moved for 3 to 5 feet then disappeared into heavier bush. Bud could hear its footsteps. The animal sounded very heavy and slow its every step sounding a loud thud. Fearing for his safety, he turned and headed, slowly, back toward the campsite looking over his shoulder every couple of seconds, afraid that the creature, whatever it was, might come after him.

Bud waited until the next morning to tell his wife. He didn't want her to be afraid throughout the night.

In 1979 in the Tay River campground Mrs. Diane Menzal and her son were walking along the highway near the river crossing. They heard a noise behind them and turned to see a large very tall hair-covered creature cross the highway, look at them, and walk up a hill on the other side of the road and into the trees.

The creature did not break into a run, it just walked like a man would walk - remaining on its hind legs the whole time. Afraid of ridicule, she kept the story to herself until hearing of a forestry worker in the area who almost hit a Sasquatch with his vehicle. Diane came forward with her story and reported the incident.

Some months later Diane's son called me to report the incident and tell his story as well. His story collaborated Diane's except for his failure to mention the creature turning to look at them before it walked off.

At the date of writing, the forestry worker has not been traced to further confirm the sighting.

CHAPTER ELEVEN
"Additional Sightings"

A man from Drayton Valley, Alberta wrote John Green with a reported sighting in the Lake Ribbon Creek area in April, 1969.

> It was around noon time that the three men built a campfire to heat up and eat some food they were carrying with them. They sat and ate and talked for about 30 minutes when one of the men said "Hey what's that?". His two companions looked to where he pointed and saw about 100 yards away from them what they later described as a gorilla sitting on his haunches watching them. Two of the men wanted to run back to their car but the third man whispered to them, "wait a minute". Apparently, the animal just sat there looking at the men for about five minutes, then it stood up. The creature then made a chattering noise with its teeth and at the same time moved its arms in an up and down motion. Later, after the creature had moved off, the men went to the spot where it had been sitting, but they did not find any footprints nor any other sign of the creature. Later, all the men agreed that the strange animal stood between seven and eight feet tall.

The man drew a sketch of the animal and included it with the letter. According to the sketch, the animal may have been female as it had long droopy breasts.

This sighting was in the same year as the Edmonton Journal report of Chief Joe Smallboys Indian band sightings (Chapter 1).

In August of 1968 Mr. Gerald Martin claimed that he and his family watched what might have been a Sasquatch walking along a ridge east of Highway 93, opposite the Columbia Ice Fields. He described the creature as a large, black upright figure walking along a ridge quite a distance away. The family agreed that it was too large and walking too fast to have been a man.

Guy Phillips, a researcher in Winnipeg, Manitoba wrote me about a report from the Lake Louise area in December 21, 1984. Guy has been looking into reports for many years and when he hears of something in the Alberta area, he will usually correspond the details to me.

Guy's letter regarding this incident follows:

"As I mentioned on the phone, I have some info on a sighting out your way from 1984. The details are very sketchy but you are welcome to them. I wrote little down at the time at it wasn't local and I relayed it verbally the same day to someone who passed it on to (Rene) Dahinden.

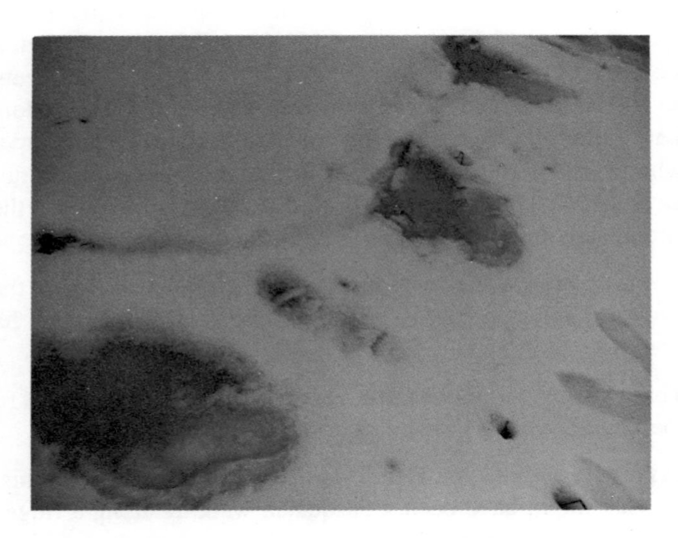

One of the photos taken by Sharon Smith (not her neal name) of the tracks she and her husband thought might be Sasquatch tracks. But I am sure they are just footprints melted out.

It took place (between) December 18 and 21, 1984 (she wasn't sure of the exact date). Some university students from Winnipeg were on a ski vacation. At one point they were driving just outside of Lake Louise when they saw an animal climbing up a mountain. It was eight to nine feet tall, a brown (dark) or black color. They (3 or 4 in a car?) stopped the car and watched it for 5 or 10 minutes. It never went on all fours and continued up the mountain side. They later returned to the chalet they were staying at and were overheard talking about their sighting by a man there. He told them he had been hiking in the same area several months before and had found and photographed some 17 inch tracks he had seen in the mud. I spoke to the girl who saw it, in August of 1985 and I think she believed she saw something that was not a bear. This incident took place at dusk, however, and using local reference points she put the distance at approximately one or two blocks. She had no idea who the man at the chalet was. They were not drinking at the time of the sighting."

Unfortunately, the report did not make it to Alberta until just recently. Therefore, the story can not be properly investigated and the man who allegedly photographed the tracks has not been identified. Also, no reports of these pictures have reached my files.

On the night of March 21, 1988 Dan and Julia (alias) reported to me their sighting of what they believe was the Sasquatch.

In May, 1983 on the Stoney Indian Reserve along Highway 1A crossing Oldfort Creek, Dan and Julia were driving over the bridge when they spotted a dark brown human-like creature step onto the road. The couple immediately identified the creature as a Sasquatch.

The creature turned to look at the oncoming car then bolted across the road, jumped a four foot fence and ran into the trees. Dan wanted to stop the car and chase after it but Julia, who was pregnant at the time, became very frightened and begged him not to go after it.

"Just keep driving" she pleaded, and Dan did just that.

The couple saw the creature at a very close proximity. It was covered with hair except for parts of the face which were covered with black skin. Dan described the eyes as large and deep set with the forehead jutting out. He claims that it stood between seven and eight feet tall with large and very muscular arms. Both the husband and wife agreed that it was remarkably fast and bolted over the fence with very little effort. When they drove past the spot on the road where the creature had jumped over the fence and into the trees, they could see the trees and brush moving as it ran deeper and deeper into the woods.

Track from 1986 siting, Chilliwack River, B.C.

Track from 1986 siting, Chilliwack River, B.C.

When Dan and Julia first reported the incident to me, they were both willing to have their names and place of residence published but later withdrew because they "wanted to put the incident behind them and go on".

Dan is a hunter however, and expressed a desire to hunt the creature.

Not all Sasquatch incidents reported are authentic. Many actually check out to be false. Occasionally I receive reports by people who sincerely

The 4 foot fence which the Sasquatch cleared with no effort at all, according to Dan and Julia.

believe they have found something which relates to the Sasquatch but the evidence is merely a case of mistaken identity.

On the night of January 26, 1987 a woman called to report a track find in the snow on a frozen swamp area not far from Rocky Mountain House, discovered in 1986. Her and her husband photographed the strange tracks and wanted to present them to me for verification.

There were six photos in all and they did indeed show large impressions almost walking in a straight line across the ice. The size of the prints were huge when compared with Sharon's 7 1/2 size running shoe along side in the photograph. The couple did not measure the size, however, or the stride. Upon close examination I concluded that the prints were not that of a Sasquatch but of a human — the prints merely melted out in warm weather.

The outline of a boot print inside of each track could be seen. That was probably the original size of the print but the warmth and the rays of the sun on the ice seemed to have melted them causing the larger impression that convinced the couple that the tracks may have belonged to a Sasquatch. Still a further clue in examining the photos was the size of the gait or the distance between the tracks. If the tracks were their original size, whatever made them was walking heel to toe at the time.

The courage of this couple to come forward and the intelligence to take photos was encouraging though. I thanked them for calling and expressed my desire to have more people come forward with possible track findings.

The bridge over the Oldfort Creek on which Dan and Julia encountered a Sasquatch in early May 1983.

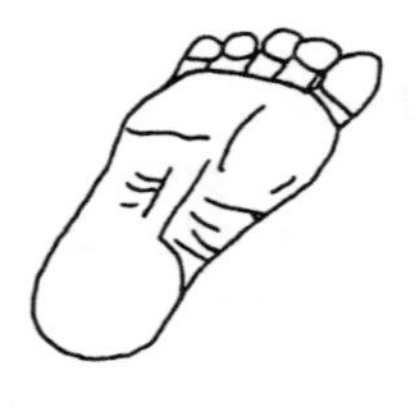
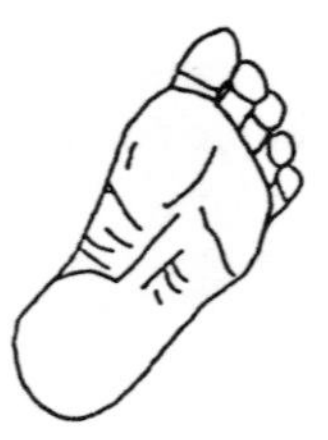
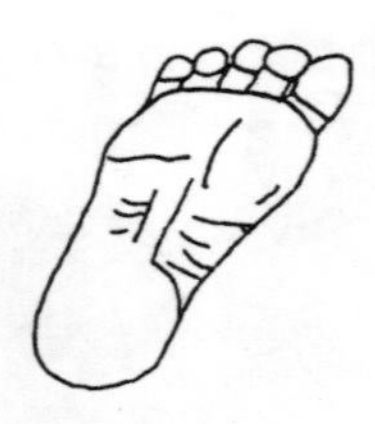
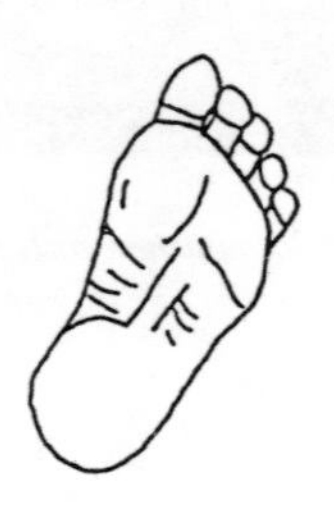

CONCLUSION
"To Shoot Or Not To Shoot"

There have been many self-proclaimed experts in the field of Sasquatch research. Some have thought to know the migration routes, caves which they occupy, rivers they seem to favor but in the opinion of this writer, there are, as yet, no experts.

Many excellent researchers are presently working towards finding a Sasquatch and proving its existence (some of which are quoted and give attribute in this book). Until, however, one of these animals is brought into captivity and examined closely and thoroughly, there will be no experts.

A popular debate among those who believe in the existence of the animal is whether or not to shoot a Sasquatch in order to prove it as a legitimate animal or breed of animal. Now, before every naturalist and environmentalist raises arms against me and threaten flogging, I will make it clear that shooting one of these creatures is not an idea I personally relish. In fact, before 1980 I was dead set against it always hoping that myself or someone else would be lucky enough to come across the remains of one who died of natural causes.

Unfortunately, dead animal flesh does not last long in the wild. Scavenger animals and birds will most often destroy a corpse before man lays his eyes upon it. Also, it is a rare occurrence to find a whole skeleton in the wild in one piece.

When a handful of men began seriously investigating the question of this animal they met with closed minds in the scientific community whose overwhelming attitude in the late 50's and all throughout the 60's was that if such a thing existed, they would surely be aware of it. Although science has witnessed some advancement in this area, the attitude remains very much the same. Most scientists claim neutrality on the subject of the Sasquatch.

"I will keep an open mind if you can present me the body of one of these creatures" closely reflects the current thinking. Very few scientists are willing to conclusively examine current evidence although many will suggest that they are convinced that there is truly something out there.

Professor Grover Krantz has studied the anatomy suggested by the footprints. He has concluded that in at least some cases the tracks in question could not have been faked. In fact, a set of footprints that Professor Krantz studied, found in Umatilla National Forest on the border between Washington and Oregon states in 1982, were clear enough to show skin marks known as dermal ridges (like finger prints on the bottom of the feet). Most finger print experts who have examined these tracks concluded that they were real. My

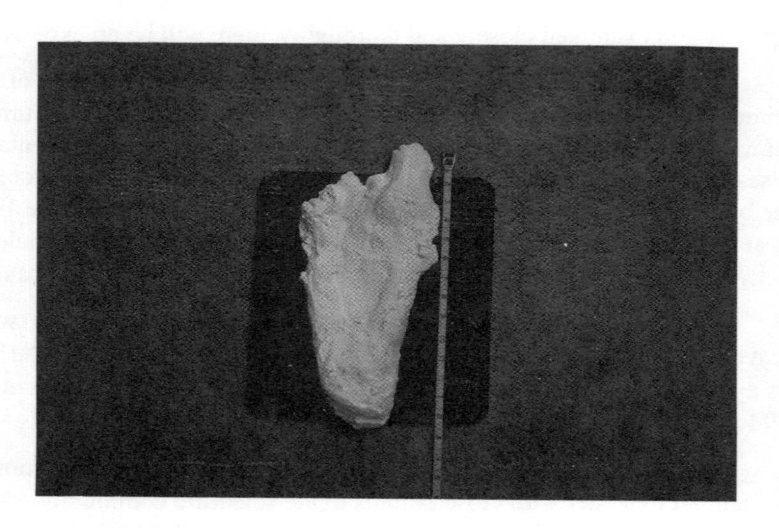

Cast of Alleged Dermal Ridge Footprints, Umatilla National Forest, 1982 (right foot).

Cast of Alleged Dermal Ridge Footprints, Umatilla National Forest, 1982 (left foot).

natural reaction was that this would lead to a serious confirmation by the scientific community that the Sasquatch is a definite possibility. It is unfortunate however that I must now share my doubts. Because the individual who first found the tracks has openly admitted to faking prints in the past, we cannot be one hundred percent sure that they had no hand in faking the dermal ridges. It has also come to my attention that given the right soil conditions etc. it is quite possible to fake dermal ridges.

Although the dermal ridges turned a few heads, little changed in the way of attitudes towards further research. Realistically speaking, the objectivity of science would demand that this information merely be filed and held for comparison with further evidence. It is this realization that has turned me around in my "shooting" question.

I am persuaded that at least one must be shot in order to prove its existence and bring it into the historical, environmental, and scientific realm. Without proof of its existence the animal cannot even be properly protected from possible extinction.

One notably different opinion on the "shoot or not shoot" question is that of my friend Professor Vladimir Markotic (Archaeology, University of Calgary) who has been avidly involved in researching the Sasquatch question since the early 1960's. Vladimir is unique in that he is the only legitimate scientist I know of who is against the shooting of a Sasquatch to prove its existence. He said to me some time ago "Tom, why should we have to prove anything anyway. If you shoot a Sasquatch and say here is the proof, the scientists who said there was no such thing will claim that they knew of it all along". No doubt Vladimir has a point here. He also said "We should not look at the subject so black and white. If the Sasquatch is real, then it is an important scientific discovery, if it is not real then it has to be looked into anyway because it is an important piece of folklore."

Vladimir has given new food for thought lure. I must admit that the only thing I was thinking of was whether or not the Sasquatch was real. So is the Sasquatch real? Is there a large bipedal animal running around in our mountains, forests and bushes?

This is for you to decide. My personal opinion is that the animal does exist. To date, there has been a total of 3000 reported incidents and the number grows by the day. Many of these reports were made by reputable people, who do not personally know others who have made the same claim and told of the same encounter.

There is also the Patterson film, the alleged dermal ridges, many tracks and clear prints, eyewitness testimonies, historical sightings from years ago, Indian Legends which tell similar stories of a similar creature. I am, however, willing to be convinced that there is nothing out there if the evidence eventually convinces me.

The question of the Sasquatch will rage on as one of the most debated subjects of this century until one is either shot or brought into captivity.

Food for thought: An unsolved mystery; a hoax; or an Indian legend that lives on.

Exciting HANCOCK HOUSE Titles

On the Track of the SASQUATCH
Encounters with BIGFOOT
from California to Canada Book 1
John Green
8½ x 11 64 pp. SC

Sasquatch
The Apes Among Us
John Green
5½ x 8½ 492 pp. SC
ISBN 0-88839-123-4

On the Track of the SASQUATCH
Encounters with BIGFOOT
from California to Canada Book 2
John Green
8½ x 11 64 pp. SC

Ogopogo
The True Story of the Okanagan Lake Million Dollar Monster
Arlene Gaal
5½ x 8½ 128 pp. SC
ISBN 0-88839-987-1

Recent sightings of the monster continue to intrigue. Many of the famous photos of Ogopogo taken over the years are in this book.

Monster hunters from around the world will not rest until the mystery is solved. Discover the fascination!

HANCOCK HOUSE Titles

Native Titles

Ah Mo
Tren J. Griffin
ISBN 0-88839-244-3

American Indian Pottery
Sharon Wirt
ISBN 0-88839-134-X

Argillite: Art of the Haida
Drew & Wilson
ISBN 0-88839-037-8

Artifacts of the NW Coast Indians
Hilary Stewart
ISBN 0-88839-098-X

Art of the Totem
Marius Barbeau
ISBN 0-88839-168-4

Coast Salish
Reg Ashwell
ISBN 0-88839-009-2

End of Custer
Dale T. Schoenberger
ISBN 0-88839-288-5

Eskimo Life Yesterday
Hancock House
ISBN 0-919654-73-8

Guide to Indian Quillworking
Christy Ann Hensler
ISBN 0-88839-214-1

Haida: Their Art & Culture
Leslie Drew
ISBN 0-88839-132-3

Hunter Series
R. Stephen Irwin, MD, Illustrations J. B. Clemens:

- **Hunters of the Buffalo**
  ISBN 0-88839-176-5
- **Hunters of the E. Forest**
  ISBN 0-88839-178-1
- **Hunters of the Ice**
  ISBN 0-88839-179-X
- **Hunters of the N. Forest**
  ISBN 0-88839-175-7
- **Hunters of the Sea**
  ISBN 0-88839-177-3

Images: Stone: B.C.
Wilson Duff
ISBN 0-295-95421-3

Incredible Eskimo
de Coccola & King
ISBN 0-88839-189-7

Indian Art & Culture
Kew & Goddard
ISBN 0-919654-13-4

Indian Artifacts of the NE
Roger W. Moeller
ISBN 0-88839-127-7

Indian Coloring Books
Carol Batdorf

- **Gifts of the Seasons**
  ISBN 0-88839-246-X
- **Seawolf**
  ISBN 0-88839-247-8
- **Tinka**
  ISBN 0-88839-249-4
- **Totem Poles**
  ISBN 0-88839-248-6

Indian Healing
Wolfgang G. Jilek, MD
ISBN 0-88839-120-X

Indian Herbs
Dr. Raymond Stark
ISBN 0-88839-077-7

Indian Rock Carvings of the Pacific NW Coast
Beth Hill
ISBN 0-919654-34-7

Indian Tribes of the NW
Reg Ashwell
ISBN 0-919654-53-3

Indian Weaving, Knitting & Basketry of the NW
Elizabeth Hawkins
ISBN 0-88839-006-8

Indians of the NW Coast
D. Allen
ISBN 0-919654-82-7

Iroquois: Their Art & Crafts
Carrie A. Lyford
ISBN 0-88839-135-8

Kwakiutl Art & Culture
Reg Ashwell
ISBN 0-88839-325-3

Kwakiutl Legends
Chief Wallas & Whitaker
ISBN 0-88839-094-7

Life with the Eskimo
Hancock House
ISBN 0-919654-72-X

More Ah Mo
Tren J. Griffin
ISBN 0-88839-303-2

My Heart Soars
Chief Dan George
ISBN 0-88839-231-1

My Spirit Soars
Chief Dan George
ISBN 0-88839-233-8

North American Indians
Mike Roberts
ISBN 0-88839-000-9

NW Native Harvest
Carol Batdorf
ISBN 0-88839-245-1

Power Quest
Carol Batdorf
ISBN 0-88839-240-0

River of Tears
Maud Emery
ISBN 0-88839-276-1

Spirit Quest
Carol Batdorf
ISBN 0-88839-210-9

Talking Stick
Carol Batdorf
ISBN 0-88839-308-3

Those Born at Koona
John & Carolyn Smyly
ISBN 0-88839-101-3

Tlingit: Art, Culture & Legends
Dan & Nan Kaiper
ISBN 0-88839-010-6

Totem Poles of the NW
D. Allen
ISBN 0-919654-83-5

When Buffalo Ran
George Bird Grinnell
ISBN 0-88839-258-3